International Mathematical Olympiads

1986–1999

Geometric diagrams by Ireneusz Świątkiewicz

Library of Congress Catalog Card Number 2003103065

ISBN 0-88385-811-8

Printed in the United States of America

Current Printing (last digit):
10 9 8 7 6 5 4 3 2 1

International Mathematical Olympiads

1986–1999

Compiled and with solutions by

Marcin E. Kuczma
University of Warsaw, Poland

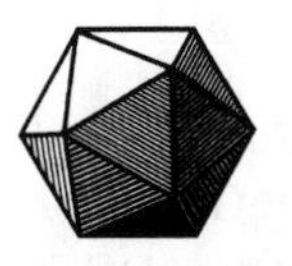

Published and Distributed by
The Mathematical Association of America

MAA PROBLEM BOOKS SERIES

Problem Books is a series of the Mathematical Association of America consisting of collections of problems and solutions from annual mathematical competitions; compilations of problems (including unsolved problems) specific to particular branches of mathematics; books on the art and practice of problem solving, etc.

A Friendly Mathematics Competition: 35 Years of Teamwork in Indiana, edited by Rick Gillman

The Inquisitive Problem Solver, Paul Vaderlind, Richard K. Guy, and Loren L. Larson

International Mathematical Olympiads 1986–1999, compiled and with solutions by Marcin E. Kuczma

Mathematical Olympiads 1998–1999: Problems and Solutions From Around the World, edited by Titu Andreescu and Zuming Feng

Mathematical Olympiads 1999–2000: Problems and Solutions From Around the World, edited by Titu Andreescu and Zuming Feng

The William Lowell Putnam Mathematical Competition 1985–2000: Problems, Solutions, and Commentary, Kiran S. Kedlaya, Bjorn Poonen, Ravi Vakil

USA and International Mathematical Olympiads 2000, edited by Titu Andreescu and Zuming Feng

USA and International Mathematical Olympiads 2001, edited by Titu Andreescu and Zuming Feng

MAA Service Center
P. O. Box 91112
Washington, DC 20090-1112
1-800-331-1622 fax: 1-301-206-9789
www.maa.org

Contents

Preface

There is little need to introduce the International Mathematical Olympiad to the readers. The IMO is the most prestigious of all international competitions in high school mathematics. It has been conducted annually since 1959 (with the single exception of the year 1980). In contrast to physical sports, the game of the IMO is to challenge the mental skills of its participants. Like a game, there are precise regulations, scores, results, and medals. There are those who win medals and those who, while losing in the narrow sense, gain immeasurably from the intensity of the effort and the competition. Several hundred incredibly gifted school students from several dozen countries compete each year, taking a two-day examination consisting of three truly difficult problems to be solved each day.

This book focuses on the mathematics of the IMO, leaving the sporting aspects of the event to the media. It is a continuation of the two previous MAA publications that comprised the first twenty-six contests: *International Mathematical Olympiads 1959–1977*, NML vol. 27, by Samuel L. Greitzer, and *International Mathematical Olympiads 1979–1985 and forty supplementary problems*, NML vol. 31, by Murray S. Klamkin. Accordingly, this volume takes up the series, starting from the 27th IMO in Poland (incidentally, the author's native country) and ending up with the 40th IMO in Romania.

The reader will find all the problems of those fourteen IMOs compiled here. There is true mathematical beauty in many of them. I have had the good luck to be present at the IMO in various roles: many years ago, as a contestant; then, several times, as the delegation leader of my country, hence a Jury member; and also, four times, as a member of the Problem Selection Committee. So I have had the chance to witness the process of selection, preliminary and ultimate, of the contest problems. Each time, the work on the problem proposals coming from the participating countries was a fascinating intellectual adventure. In so many cases, we could only regret that there was room for no more than six contest problems, so that one or another tiny mathematical treasure had to be left out.

Each problem in this book is given with a solution, sometimes more than one. Occasionally it is accompanied by a remark, in which an alternative approach or a generalization, or the relevance of the problem to some mathematical theory, is mentioned. Sometimes the remark is intended as a useful tool "from the tool-chest": a brief exposition of a solution technique that goes slightly beyond the usual school curriculum.

In many cases the solutions are patterned on the proposers' solution sketches. Many other ones have been inspired by the contestants' papers; it is not at all rare at olympiads and contests that credit for the trickiest and smartest methods goes to the students. Which means that the problems in this volume, as well as their solutions, have been created or inspired by many people, whom I could in no way be able to name here. They have to remain anonymous; but it must be remembered that they are, in fact, the true authors of this book. My gratitude to them is most sincere.

Although the solutions are due to several authors, the exposition is mine; and of course, the responsibility is mine for all mistakes that may have appeared. Needless to say, I would never dare claim to have devised the nicest solutions available. The readers who study this book, be it for training or just for recreation, will certainly find yet other ways of resolving this or that problem, possibly more elegant or more general than the presented ones. So much the better! Satisfaction from a good job done is the solver's true reward.

I will be grateful for any comments from the readers, for communication about flaws noticed, about other nice solutions and possible improvements to those given here. I wish all the readers joy, fun and pleasure in tackling the problems.

Marcin E. Kuczma
Institute of Mathematics
University of Warsaw
ul. Banacha 2, PL 02097
Warsaw, Poland

How the Book Is Organized

As mentioned in the *Preface*, the book is a sequel to two earlier MAA publications of past IMO competitions; it is therefore similar in form to those books. The *Problems* section comes first, containing the statements of all 84 problems, in basically the same wording in which they were posed at the IMOs. The numbering is by the IMO year and, within each particular IMO, by a number from 1 to 6; the book begins with problem 1986/1 (the first question at the 1986 IMO) and ends with problem 1999/6.

In the *Solutions* section the problem statements are not repeated. The proposed solutions are rather detailed, especially the first solutions of problems to which two or more solutions are given. Some displayed formulas are numbered for the sake of further reference; their numbering starts anew with each subsequent problem (but it continues through the second and third solution to the same problem).

In the *Results* section the reader will find tables that, traditionally, present a summary of prizes and medals and the joint scores of contestants from participating countries. It must be, however, remembered that the IMO is an individual competition and that any countries ranking has an unofficial character.

The *List of Symbols* and *Glossary* come next. Terms and theorems that exceed the standard school syllabus, arranged in alphabetic order, are briefly explained. They involve no advanced mathematics, just some concepts and facts, used in this book, with which every student aiming at olympiads and contests should be acquainted. The explanations are meant as a reminder, and not a source for learning a theory.

The *Subject Classification* is an attempt to subject classification. This can be problematic, since problems that combine ideas from various fields of mathematics are very popular at IMOs. The reader may disagree with what is considered by the author as the "primary classification" of this or that problem.

Bibliography is the concluding section. There exists a vast literature on math contests, problem-solving techniques, and contest problem compilations elaborated by many authors from various countries; and most of these are excellent elaborations. Books that might be recommended are counted in the hundreds. Thus the bibliographical items included here are to be considered as a selection from a much broader offering—a proposal for possible further reading.

Problems

Twenty-seventh International Olympiad, 1986

1986/1. Let d be any positive integer not equal to 2, 5 or 13. Show that one can find distinct a, b in the set $\{2, 5, 13, d\}$ such that $ab - 1$ is not a perfect square.

1986/2. A triangle $A_1A_2A_3$ and a point P_0 are given in the plane. We define $A_s = A_{s-3}$ for all $s \geq 4$. We construct a sequence of points $P_1, P_2, P_3, \ldots$ such that P_{k+1} is the image of P_k under rotation with center A_{k+1} through angle $120°$ clockwise (for $k = 0, 1, 2, \ldots$). Prove that if $P_{1986} = P_0$, then the triangle $A_1A_2A_3$ is equilateral.

1986/3. To each vertex of a regular pentagon an integer is assigned in such a way that the sum of all the five numbers is positive. If three consecutive vertices are assigned the numbers x, y, z respectively and $y < 0$, then the following operation is allowed: the numbers x, y, z are replaced by $x + y$, $-y$, $z + y$ respectively.

Such an operation is performed repeatedly as long as at least one of the five numbers is negative. Determine whether this procedure necessarily comes to an end after a finite number of steps.

1986/4. Let A, B be adjacent vertices of a regular n-gon ($n \geq 5$) in the plane having center at O. A triangle XYZ, which is congruent to and initially coincides with OAB, moves in the plane in such a way that Y and Z each trace out the whole boundary of the polygon, X remaining inside the polygon. Find the locus of X.

1986/5. Find all functions f, defined on the nonnegative real numbers and taking nonnegative real values, such that:

(i) $f(xf(y))f(y) = f(x + y)$ for all $x, y \geq 0$,

(ii) $f(2) = 0$,

(iii) $f(x) \neq 0$ for $0 \leq x < 2$.

1986/6. One is given a finite set of points in the plane, each point having integer coordinates. Is it always possible to color some of the points in the set red and the remaining points white in such a way that for any straight line L parallel to either one of the coordinate axes the difference (in absolute value) between the numbers of white points and red points on L is not greater than 1? Justify your answer.

Twenty-eighth International Olympiad, 1987

1987/1. Let $p_n(k)$ be the number of those permutations of the set $\{1, \ldots, n\}$ that have exactly k fixed points. Prove that

$$\sum_{k=0}^{n} k \cdot p_n(k) = n!.$$

(Remark: A permutation f of a set S is a one-to-one mapping of S onto itself. An element i in S is called a fixed point of the permutation f if $f(i) = i$.)

1987/2. In an acute-angled triangle ABC the interior bisector of the angle A intersects BC at L and intersects the circumcircle of ABC again at N. From point L perpendiculars are drawn to AB and AC, the feet of these perpendiculars being K and M respectively. Prove that the quadrilateral $AKNM$ and the triangle ABC have equal area.

1987/3. Let $x_1, x_2, \ldots, x_n$ be real numbers satisfying the equation

$$x_1^2 + x_2^2 + \cdots + x_n^2 = 1.$$

Prove that for every integer $k \geq 2$ there are integers $a_1, a_2, \ldots, a_n$, not all 0, such that $|a_i| \leq k - 1$ for all i and

$$|a_1x_1 + a_2x_2 + \cdots + a_nx_n| \leq \frac{(k-1)\sqrt{n}}{k^n - 1}.$$

1987/4. Prove that there is no function f from the set of nonnegative integers into itself such that $f(f(n)) = n + 1987$ for every n.

1987/5. Let n be an integer greater than or equal to 3. Prove that there is a set of n points in the plane such that the distance between any two points is irrational and each set of three points determines a nondegenerate triangle with rational area.

1987/6. Let n be an integer greater than or equal to 2. Prove that if the number $k^2 + k + n$ is prime for all integers k such that $0 \le k \le \sqrt{n/3}$, then $k^2 + k + n$ is prime for all integers k such that $0 \le k \le n - 2$.

Twenty-ninth International Olympiad, 1988

1988/1. Consider two coplanar circles of radii R and r $(R > r)$ with the same center. Let P be a fixed point on the smaller circle and B a variable point on the larger circle. The line BP meets the larger circle again at C. The perpendicular ℓ to BP at P meets the smaller circle again at A (if ℓ is tangent to the circle at P, then $A = P$).

(i) Find the set of values of $BC^2 + CA^2 + AB^2$.

(ii) Find the locus of the midpoint of AB.

1988/2. Let n be a positive integer and let $A_1, A_2, \ldots, A_{2n+1}$ be subsets of a set B. Suppose that

(a) each A_i has exactly $2n$ elements,

(b) each

$$A_i \cap A_j \ (1 \le i < j \le 2n + 1)$$

contains exactly one element,

(c) every element of B belongs to at least two of the A_i.

For which values of n can one assign to every element of B one of the numbers 0 and 1 in such a way that each A_i has 0 assigned to exactly n of its elements?

1988/3. A function f is defined on the positive integers by

$$f(1) = 1, \quad f(3) = 3, \quad f(2n) = f(n),$$
$$f(4n + 1) = 2f(2n + 1) - f(n),$$
$$f(4n + 3) = 3f(2n + 1) - 2f(n),$$

for all positive integers n. Determine the number of positive integers n, less than or equal to 1988, for which $f(n) = n$.

1988/4. Show that the set of real numbers x that satisfy the inequality

$$\sum_{k=1}^{70} \frac{k}{x-k} \geq \frac{5}{4}$$

is a union of disjoint intervals, the sum of whose lengths is 1988.

1988/5. ABC is a triangle right-angled at A, and D is the foot of the altitude from A. The straight line joining the incenters of the triangles ABD, ACD intersects the sides AB, AC at the points K, L respectively. S and T denote the areas of the triangles ABC and AKL respectively. Show that $S \geq 2T$.

1988/6. Let a and b be positive integers such that $ab+1$ divides a^2+b^2. Show that

$$\frac{a^2+b^2}{ab+1}$$

is the square of an integer.

Thirtieth International Olympiad, 1989

1989/1. Prove that the set $\{1, 2, \ldots, 1989\}$ can be expressed as the disjoint union of subsets A_i $(i = 1, 2, \ldots, 117)$ such that

(i) each A_i contains 17 elements;

(ii) the sum of all the elements in each A_i is the same.

1989/2. In an acute-angled triangle ABC the internal bisector of angle A meets the circumcircle of the triangle again at A_1. Points B_1 and C_1 are defined similarly. Let A_0 be the point of intersection of the line AA_1 with the external bisectors of angles B and C. Points B_0 and C_0 are defined similarly. Prove that

(a) the area of the triangle $A_0B_0C_0$ is twice the area of the hexagon $AC_1BA_1CB_1$;

(b) the area of the triangle $A_0B_0C_0$ is at least four times the area of the triangle ABC.

1989/3. Let n and k be positive integers and let S be a set of n points in the plane such that

(i) no three points of S are collinear, and

(ii) for every point P of S there are at least k points of S equidistant from P.

Prove that $k < \frac{1}{2} + \sqrt{2n}$.

1989/4. Let $ABCD$ be a convex quadrilateral such that the sides AB, AD, BC satisfy $AB = AD + BC$. There exists a point P inside the quadrilateral at a distance h from the line CD such that $AP = h + AD$ and $BP = h + BC$. Show that

$$\frac{1}{\sqrt{h}} \geq \frac{1}{\sqrt{AD}} + \frac{1}{\sqrt{BC}}.$$

1989/5. Prove that for each positive integer n there exist n consecutive positive integers none of which is an integral power of a prime number.

1989/6. A permutation $(x_1, x_2, \ldots, x_{2n})$ of the set $\{1, 2, \ldots, 2n\}$, where n is a positive integer, is said to have property P if $|x_i - x_{i+1}| = n$ for at least one i in $\{1, 2, \ldots, 2n-1\}$. Show that, for each n, there are more permutations with property P than without.

Thirty-first International Olympiad, 1990

1990/1. Two chords AB, CD of a circle intersect at a point E inside the circle. Let M be an interior point of the segment EB. The tangent line at E to the circle through D, E, M intersects the lines BC, AC at F, G respectively. If $AM/AB = t$, find EG/EF in terms of t.

1990/2. Let $n \geq 3$ and consider a set E of $2n-1$ distinct points on a circle. Suppose that exactly k of these points are to be colored black. Such a coloring is *good* if there is at least one pair of black points such that the interior of one of the arcs between them contains exactly n points from E. Find the smallest value of k so that every such coloring of k points of E is good.

1990/3. Determine all integers $n > 1$ such that

$$\frac{2^n + 1}{n^2}$$

is an integer.

1990/4. Let Q^+ be the set of positive rational numbers. Construct a function $f\colon Q^+ \to Q^+$ such that

$$f(xf(y)) = \frac{f(x)}{y} \quad \text{for all } x, y \text{ in } Q^+.$$

1990/5. Given an initial integer $n_0 > 1$, two players A and B choose integers $n_1, n_2, n_3, \ldots$ alternately according to the following rules. Knowing n_{2k}, A chooses any integer n_{2k+1} such that $n_{2k} \le n_{2k+1} \le n_{2k}^2$. Knowing n_{2k+1}, B chooses any integer n_{2k+2} such that n_{2k+1}/n_{2k+2} is a positive power of a prime. Player A wins the game by choosing the number 1990, player B wins by choosing the number 1. For which n_0 does

A have a winning strategy,
B have a winning strategy,
neither player have a winning strategy?

1990/6. Prove that there exists a convex 1990-gon with the following two properties:

(a) all angles are equal;
(b) the lengths of the sides are the numbers $1^2, 2^2, 3^2, \ldots, 1989^2, 1990^2$ in some order.

Thirty-second International Olympiad, 1991

1991/1. Given a triangle ABC, let I be the center of its inscribed circle. The internal bisectors of the angles A, B, C meet the opposite sides in A', B', C' respectively. Prove that

$$\frac{1}{4} < \frac{AI \cdot BI \cdot CI}{AA' \cdot BB' \cdot CC'} \le \frac{8}{27}.$$

1991/2. Let $n > 6$ be an integer and $a_1, a_2, \ldots, a_k$ be all the natural numbers less than n and relatively prime to n. If

$$a_2 - a_1 = a_3 - a_2 = \cdots = a_k - a_{k-1} > 0,$$

prove that n must be either a prime or a power of 2.

1991/3. Let $S = \{1, 2, \ldots, 280\}$. Find the smallest integer n such that each n-element subset of S contains five numbers that are pairwise relatively prime.

1991/4. Suppose G is a connected graph with k edges. Prove that it is possible to label the edges $1, 2, 3, \ldots, k$ in such a way that at each vertex that belongs to two or more edges, the greatest common divisor of the integers labeling those edges is equal to 1.

[A *graph* G consists of a set of points, called *vertices*, together with a set of *edges* joining certain pairs of distinct vertices. Each pair of vertices u, v belongs to at most one edge. The graph G is *connected* if for each pair of distinct vertices x, y there exists some sequence of vertices

$$x = v_0, v_1, v_2, \ldots, v_m = y$$

such that each pair v_i, v_{i+1} $(0 \leq i < m)$ is joined by an edge of G.]

1991/5. Let ABC be a triangle and P an interior point in ABC. Show that at least one of the angles $\angle PAB$, $\angle PBC$, $\angle PCA$ is less than or equal to $30°$.

1991/6. An infinite sequence $x_0, x_1, x_2, \ldots$ of real numbers is said to be *bounded* if there is a constant C such that $|x_i| \leq C$ for every $i \geq 0$.

Given any real number $a > 1$, construct a bounded infinite sequence $x_0, x_1, x_2, \ldots$ such that

$$|x_i - x_j| \cdot |i - j|^a \geq 1$$

for every pair of distinct nonnegative integers i, j.

Thirty-third International Olympiad, 1992

1992/1. Find all integers a, b, c with $1 < a < b < c$ such that the product $(a-1)(b-1)(c-1)$ is a divisor of $abc - 1$.

1992/2. Let R denote the set of all real numbers. Find all functions $f\colon \mathrm{R} \to \mathrm{R}$ such that

$$f(x^2 + f(y)) = y + (f(x))^2 \quad \text{for all } x, y \text{ in R.}$$

1992/3. Consider nine points in space, no four of which are coplanar. Each pair of points is joined by an edge (that is, a line segment) and each edge is either colored blue or red or left uncolored. Find the smallest value of n such that whenever exactly n edges are colored, the set of colored edges necessarily contains a triangle all of whose edges have the same color.

1992/4. In the plane let C be a circle, L a line tangent to the circle C and M a point on L. Find the locus of all points P with the following property: there exist two points Q, R on L such that M is the midpoint of QR and C is the inscribed circle of triangle PQR.

1992/5. Let S be a finite set of points in three-dimensional space. Let S_x, S_y, S_z be the sets consisting of the orthogonal projections of the points of S onto the yz-plane, zx-plane, xy-plane respectively. Prove that

$$|S|^2 \leq |S_x| \cdot |S_y| \cdot |S_z|$$

where $|A|$ denotes the number of elements in the finite set A. (Note: the orthogonal projection of a point onto a plane is the foot of the perpendicular from the point to the plane.)

1992/6. For each positive integer n, $S(n)$ is defined to be the greatest integer such that, for every positive integer $k \leq S(n)$, n^2 can be written as the sum of k positive square integers.

(a) Prove that $S(n) \leq n^2 - 14$ for each $n \geq 4$.

(b) Find an integer n such that $S(n) = n^2 - 14$.

(c) Prove that there exist infinitely many positive integers n such that $S(n) = n^2 - 14$.

Thirty-fourth International Olympiad, 1993

1993/1. Let $f(x) = x^n + 5x^{n-1} + 3$ where $n > 1$ is an integer. Prove that $f(x)$ cannot be expressed as the product of two polynomials, each of which has all its coefficients integers and degree at least 1.

1993/2. Let D be a point inside the acute-angled triangle ABC such that

$$\angle ADB = \angle ACB + 90^\circ \quad \text{and} \quad AC \cdot BD = AD \cdot BC.$$

(a) Calculate the value of the ratio

$$\frac{AB \cdot CD}{AC \cdot BD}.$$

(b) Prove that the tangents at C to the circumcircles of the triangles ACD and BCD are perpendicular.

1993/3. On an infinite chessboard, a game is played as follows. At the start n^2 pieces are arranged on the chessboard in an $n \times n$-block of adjoining squares, one piece in each square. A move in the game is a jump in a horizontal or vertical direction over an adjacent occupied square to an unoccupied square immediately beyond. The piece that has been jumped over is removed. Find those values of n for which the game can end with only one piece remaining on the board.

1993/4. For three points P, Q, R in the plane, we define $m(PQR)$ to be the minimum of the lengths of the altitudes of the triangle PQR (where $m(PQR) = 0$ when P, Q, R are collinear).

Let A, B, C be given points in the plane. Prove that, for any point X in the plane,

$$m(ABC) \leq m(ABX) + m(AXC) + m(XBC).$$

1993/5. Let $\mathrm{N} = \{1, 2, 3, \ldots\}$. Determine whether or not there exists a function $f\colon \mathrm{N} \to \mathrm{N}$ such that

$$\begin{aligned} f(1) &= 2, \\ f(f(n)) &= f(n) + n \quad \text{for all } n \text{ in N}, \\ \text{and} \quad f(n) &< f(n+1) \quad \text{for all } n \text{ in N}. \end{aligned}$$

1993/6. Let $n > 1$ be an integer. There are n lamps $L_0, L_1, \ldots, L_{n-1}$ arranged in a circle. Each lamp is either ON or OFF. A sequence of steps $S_0, S_1, \ldots, S_i, \ldots$ is carried out. Step S_j affects the state of L_j only (leaving the state of all other lamps unaltered) as follows:

—if L_{j-1} is ON, S_j changes the state of L_j from ON to OFF or from OFF to ON;

—if L_{j-1} is OFF, S_j leaves the state of L_j unchanged.

The lamps are labeled (mod n), that is,

$$L_{-1} = L_{n-1}, \quad L_0 = L_n, \quad L_1 = L_{n+1}, \quad \text{etc.}$$

Initially all lamps are ON. Show that:

(a) there is a positive integer $M(n)$ such that after $M(n)$ steps all the lamps are ON again;

(b) if n has the form 2^k, then all the lamps are ON after $n^2 - 1$ steps;

(c) if n has the form $2^k + 1$, then all the lamps are ON after $n^2 - n + 1$ steps.

Thirty-fifth International Olympiad, 1994

1994/1. Let m and n be positive integers. Let $a_1, a_2, \ldots, a_m$ be distinct elements of $\{1, 2, \ldots, n\}$ such that whenever $a_i + a_j \leq n$ for some i, j, $1 \leq i \leq j \leq m$, there exists k, $1 \leq k \leq m$, with $a_i + a_j = a_k$. Prove that

$$\frac{a_1 + a_2 + \cdots + a_m}{m} \geq \frac{n+1}{2}.$$

1994/2. ABC is an isosceles triangle with $AB = AC$. Suppose that

(a) M is the midpoint of BC and O is the point on the line AM such that OB is perpendicular to AB;

(b) Q is an arbitrary point on the segment BC different from B and C;

(c) E lies on the line AB and F lies on the line AC such that E, Q and F are collinear.

Prove that OQ is perpendicular to EF if and only if $QE = QF$.

1994/3. For any positive integer k, let $f(k)$ be the number of elements in the set $\{k+1, k+2, \ldots, 2k\}$ whose base 2 representation has precisely three 1s.

(a) Prove that, for each positive integer m, there exists at least one positive integer k such that $f(k) = m$.

(b) Determine all positive integers m for which there exists exactly one k with $f(k) = m$.

1994/4. Determine all ordered pairs (m, n) of positive integers such that

$$\frac{n^3 + 1}{mn - 1}$$

is an integer.

1994/5. Let S be the set of real numbers strictly greater than -1. Find all functions $f \colon S \to S$ satisfying the two conditions:

(a) $f(x + f(y) + xf(y)) = y + f(x) + yf(x)$ for all x and y in S;

(b) $f(x)/x$ is strictly increasing on each of the intervals $-1 < x < 0$ and $0 < x$.

1994/6. Show that there exists a set A of positive integers with the following property: for any infinite set S of primes there exist two positive integers $m \in A$ and $n \notin A$, each of which is a product of k distinct elements of S for some $k \geq 2$.

Thirty-sixth International Olympiad, 1995

1995/1. Let A, B, C and D be four distinct points on a line, in that order. The circles with diameters AC and BD intersect at the points X and Y. The line XY meets BC at the point Z. Let P be a point on the line XY different from Z. The line CP intersects the circle with diameter AC at the points C and M, and the line BP intersects the circle with diameter BD at the points B and N. Prove that the lines AM, DN and XY are concurrent.

1995/2. Let a, b and c be positive real numbers such that $abc = 1$. Prove that

$$\frac{1}{a^3(b+c)} + \frac{1}{b^3(c+a)} + \frac{1}{c^3(a+b)} \geq \frac{3}{2}.$$

1995/3. Determine all integers $n > 3$ for which there exist n points $A_1, A_2, \ldots, A_n$ in the plane, and real numbers $r_1, r_2, \ldots, r_n$ satisfying the following two conditions:

(i) no three of the points $A_1, A_2, \ldots, A_n$ lie on a line;

(ii) for each triple i, j, k $(1 \leq i < j < k \leq n)$ the triangle $A_iA_jA_k$ has area equal to $r_i + r_j + r_k$.

1995/4. Find the maximum value of x_0 for which there exists a sequence of positive real numbers $x_0, x_1, \ldots, x_{1995}$ satisfying the two conditions:

(i) $x_0 = x_{1995}$;

(ii) $x_{i-1} + \dfrac{2}{x_{i-1}} = 2x_i + \dfrac{1}{x_i}$ for each $i = 1, 2, \ldots, 1995$.

1995/5. Let $ABCDEF$ be a convex hexagon with

$$AB = BC = CD, \quad DE = EF = FA \quad \text{and} \quad \angle BCD = \angle EFA = 60^\circ.$$

Let G and H be two points in the interior of the hexagon such that $\angle AGB = \angle DHE = 120^\circ$. Prove that

$$AG + GB + GH + DH + HE \geq CF.$$

1995/6. Let p be an odd prime number. Find the number of subsets A of the set $\{1, 2, \ldots, 2p\}$ such that

(i) A has exactly p elements, and

(ii) the sum of all elements in A is divisible by p.

Thirty-seventh International Olympiad, 1996

1996/1. Let $ABCD$ be a rectangular board with $AB = 20$, $BC = 12$. The board is divided into 20×12 unit squares. Let r be a given positive integer. A coin can be moved from one square to another if and only if the distance between the centers of the two squares is $\sqrt{r}$. The task is to find a sequence of moves taking the coin from the square that has A as a vertex to the square that has B as a vertex.

(a) Show that the task cannot be done if r is divisible by 2 or 3.

(b) Prove that the task can be done if $r = 73$.

(c) Can the task be done when $r = 97$?

1996/2. Let P be a point inside triangle ABC such that

$$\angle APB - \angle ACB = \angle APC - \angle ABC.$$

Let D, E be the incenters of triangles APB, APC respectively. Show that AP, BD and CE meet at a point.

1996/3. Let $N_0 = \{0, 1, 2, 3, \ldots\}$ be the set of nonnegative integers. Find all functions f defined on N_0 and taking their values in N_0 such that

$$f(m + f(n)) = f(f(m)) + f(n) \qquad \text{for all } m, n \text{ in } N_0.$$

1996/4. The positive integers a and b are such that the numbers $15a + 16b$ and $16a - 15b$ are both squares of positive integers. Find the least possible value that can be taken by the minimum of these two squares.

1996/5. Let $ABCDEF$ be a convex hexagon such that AB is parallel to ED, BC is parallel to FE and CD is parallel to AF. Let R_A, R_B, R_C denote the circumradii of triangles FAB, BCD, DEF respectively, and let p denote the perimeter of the hexagon. Prove that

$$R_A + R_B + R_C \geq \frac{p}{2}.$$

1996/6. Let n, p, q be positive integers with $n > p+q$. Let $x_0, x_1, \ldots, x_n$ be integers satisfying the following conditions:

(a) $x_0 = x_n = 0$;

(b) for each integer i with $1 \leq i \leq n$,

$$\text{either} \quad x_i - x_{i-1} = p \quad \text{or} \quad x_i - x_{i-1} = -q.$$

Show that there exists a pair (i, j) of indices with $i < j$ and $(i, j) \neq (0, n)$ such that $x_i = x_j$.

Thirty-eighth International Olympiad, 1997

1997/1. In the plane the points with integer coordinates are the vertices of unit squares. The squares are colored alternately black and white (as on a chessboard).

For any pair of positive integers m and n, consider a right-angled triangle whose vertices have integer coordinates and whose legs, of lengths m and n, lie along the edges of the squares. Let S_1 be the total area of the black part of the triangle and S_2 be the total area of the white part. Let $f(m,n) = |S_1 - S_2|$.

(a) Calculate $f(m,n)$ for all positive integers m and n that are either both even or both odd.

(b) Prove that $f(m,n) \le \frac{1}{2}\max\{m,n\}$ for all m and n.

(c) Show that there is no constant C such that $f(m,n) < C$ for all m and n.

1997/2. Angle A is the smallest in the triangle ABC. The points B and C divide the circumcircle of the triangle into two arcs. Let U be an interior point of the arc between B and C which does not contain A. The perpendicular bisectors of AB and AC meet the line AU at V and W, respectively. The lines BV and CW meet at T. Show that

$$AU = TB + TC.$$

1997/3. Let $x_1, x_2, \ldots, x_n$ be real numbers satisfying the conditions:

$$|x_1 + x_2 + \cdots + x_n| = 1$$

and

$$|x_i| \le \frac{n+1}{2} \quad \text{for } i = 1, 2, \ldots, n.$$

Show that there exists a permutation $y_1, y_2, \ldots, y_n$ of $x_1, x_2, \ldots, x_n$ such that

$$|y_1 + 2y_2 + \cdots + ny_n| \le \frac{n+1}{2}.$$

1997/4. An $n \times n$ matrix (square array) whose entries come from the set $S = \{1, 2, \ldots, 2n-1\}$ is called a *silver* matrix if, for each $i = 1, \ldots, n$, the ith row and the ith column together contain all elements of S. Show that

(a) there is no silver matrix for $n = 1997$;

(b) silver matrices exist for infinitely many values of n.

1997/5. Find all pairs (a, b) of integers $a \geq 1, b \geq 1$ that satisfy the equation

$$a^{b^2} = b^a.$$

1997/6. For each positive integer n, let $f(n)$ denote the number of ways of representing n as a sum of powers of 2 with nonnegative integer exponents. Representations that differ only in the ordering of their summands are considered to be the same. For instance, $f(4) = 4$ because the number 4 can be represented in the following four ways: 4; $2+2$; $2+1+1$; $1+1+1+1$. Prove that, for any integer $n \geq 3$,

$$2^{n^2/4} < f(2^n) < 2^{n^2/2}.$$

Thirty-ninth International Olympiad, 1998

1998/1. In the convex quadrilateral $ABCD$, the diagonals AC and BD are perpendicular and the opposite sides AB and DC are not parallel. Suppose that the point P, where the perpendicular bisectors of AB and DC meet, is inside $ABCD$. Prove that $ABCD$ is a cyclic quadrilateral if and only if the triangles ABP and CDP have equal areas.

1998/2. In a competition, there are a contestants and b examiners, where $b \geq 3$ is an odd integer. Each examiner rates each contestant as either "pass" or "fail." Suppose k is a number such that, for any two examiners, their ratings coincide for at most k contestants. Prove that

$$\frac{k}{a} \geq \frac{b-1}{2b}.$$

1998/3. For any positive integer n, let $d(n)$ denote the number of positive divisors of n (including 1 and n itself). Determine all positive integers k such that

$$\frac{d(n^2)}{d(n)} = k$$

for some n.

1998/4. Determine all pairs (a, b) of positive integers such that ab^2+b+7 divides $a^2b + a + b$.

1998/5. Let I be the incenter of triangle ABC. Let the incircle of ABC touch the sides BC, CA and AB at K, L and M, respectively. The line through B parallel to MK meets the lines LM and LK at R and S, respectively. Prove that $\angle RIS$ is acute.

1998/6. Consider all functions f from the set N of all positive integers into itself satisfying the equation

$$f(t^2 f(s)) = s(f(t))^2,$$

for all s and t in N. Determine the least possible value of $f(1998)$.

Fortieth International Olympiad, 1999

1999/1. Determine all finite sets S of at least three points in the plane that satisfy the following condition: for any two distinct points A and B in S, the perpendicular bisector of the line segment AB is an axis of symmetry for S.

1999/2. Let n be a fixed integer, with $n \geq 2$.

(a) Determine the least constant C such that the inequality

$$\sum_{1 \leq i < j \leq n} x_i x_j (x_i^2 + x_j^2) \leq C \left(\sum_{1 \leq i \leq n} x_i \right)^4$$

holds for all real numbers $x_1, \ldots, x_n \geq 0$.

(b) For this constant C, determine when equality holds.

1999/3. Consider an $n \times n$ square board, where n is a fixed even positive integer. The board is divided into n^2 unit squares. We say that two different squares on the board are *adjacent* if they have a common side.

N unit squares on the board are marked in such a way that every square (marked or unmarked) on the board is adjacent to at least one marked square. Determine the smallest possible value of N.

1999/4. Determine all pairs (n, p) of positive integers such that p is a prime, $n \leq 2p$, and $(p-1)^n + 1$ is divisible by n^{p-1}.

1999/5. Two circles Γ_1 and Γ_2 are contained inside the circle Γ, and are tangent to Γ at the distinct points M and N, respectively. Γ_1 passes through the center of Γ_2. The line passing through the two points of intersection of Γ_1 and Γ_2 meets Γ at A and B. The lines MA and MB meet Γ_1 at C and D, respectively. Prove that CD is tangent to Γ_2.

1999/6. Determine all functions $f\colon \mathrm{R} \to \mathrm{R}$ such that

$$f(x - f(y)) = f(f(y)) + xf(y) + f(x) - 1$$

for all $x, y \in \mathrm{R}$.

Solutions

Twenty-seventh International Olympiad, 1986

1986/1

It will be enough to prove that at least one of the numbers $2d - 1$, $5d - 1$, $13d - 1$ is not a perfect square. Assume, by the way of contradiction, that $2d = x^2 + 1$, $5d = y^2 + 1$, $13d = z^2 + 1$ for some integers x, y, z. By the first equation, x must be odd: $x = 2t + 1$. Hence also the number $d = (x^2 + 1)/2 = 2t^2 + 2t + 1$ is odd, and consequently y and z are even: $y = 2u$, $z = 2v$.

From $z^2 - y^2 = 8d$ we get $(v + u)(v - u) = 2d$. The number $2d$ is divisible by 2 but not by 4. This gives the desired contradiction because the numbers $v + u$ and $v - u$ are of the same parity.

1986/2

First Solution. Consider the transformation of the plane defined as the composition $f = r_3 \circ r_2 \circ r_1$, where r_j is the rotation about A_j through $120°$ clockwise. The mapping f preserves the length and direction of every vector; hence it is the translation by a certain vector $\mathbf{v}$. By the conditions of the problem, $f(P_0) = P_3$, and by periodicity, $f^n(P_0) = P_{3n}$; the symbol f^n denotes the n-fold composition (iterate) $f \circ f \circ \cdots \circ f$. Since f is the translation by $\mathbf{v}$, f^n is the translation by the vector $n\mathbf{v}$. For $n = 662$ we get $P_{3n} = P_{1986} = P_0$. Thus $\mathbf{v}$ is the zero vector, which means that f is the identity map. Denote the point $r_2(A_1)$ by B. Then $A_1 = f(A_1) = r_3(r_2(r_1(A_1))) = r_3(r_2(A_1))) = r_3(B)$.

The isosceles triangles A_1A_2B and BA_3A_1 do not coincide; they have equal angles ($\angle A_2 = \angle A_3 = 120°$) and a common base A_1B, hence they are congruent. Consequently $A_1A_2BA_3$ is a rhombus (with angles 60° and 120°), and hence $A_1A_2A_3$ is an equilateral triangle.

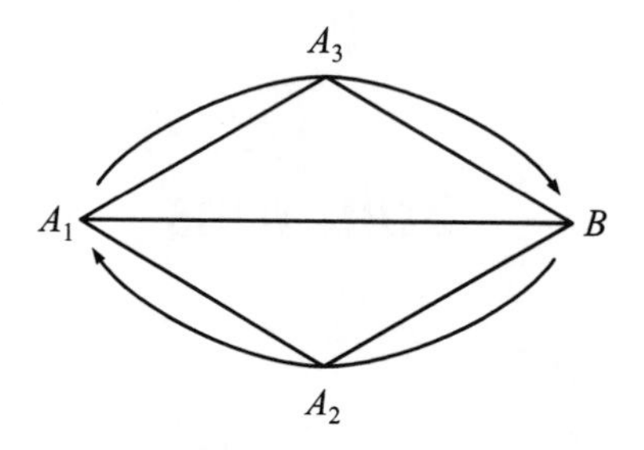

Second Solution. The same reasoning can be neatly written down in terms of the complex plane. The given points A_k, P_k ($k = 0, 1, 2, \ldots$) are represented by the complex numbers a_k, p_k. Let the symbols r_1, r_2, r_3 and f denote the same mappings as in the first solution. The rotation r_k takes any point (complex number) z to its image $r_k(z) = (z - a_k)\lambda + a_k$ where λ is the cubic root of unity

$$\lambda = \cos\frac{2}{3}\pi - i\sin\frac{2}{3}\pi = -\frac{1}{2}(1 + \sqrt{3}i).$$

Since $\lambda^3 = 1$,

$$\begin{aligned} f(z) &= (((z - a_1)\lambda + a_1 - a_2)\lambda + a_2 - a_3)\,\lambda + a_3 \\ &= (z - a_1)\lambda^3 + (a_1 - a_2)\lambda^2 + (a_2 - a_3)\lambda + a_3 = z + \omega, \end{aligned}$$

where

$$\begin{aligned} \omega &= (a_1 - a_2)\lambda^2 + (a_2 - a_3)\lambda + (a_3 - a_1) \\ &= (\lambda - 1)\,((a_1 - a_2)(\lambda + 1) + (a_2 - a_3)). \end{aligned}$$

By assumption, $p_0 = p_{1986} = f^{662}(p_0) = p_0 + 662 \cdot \omega$, showing that $\omega = 0$; and since $\lambda - 1 \neq 0$, we obtain $(a_1 - a_2)(\lambda + 1) + (a_2 - a_3) = 0$. Note that $a_1 - a_2 \neq 0$ (the points a_1, a_2, a_3 are the vertices of a triangle). Division by $a_1 - a_2$ now yields

$$\frac{a_3 - a_2}{a_1 - a_2} = \lambda + 1 = \frac{1 - \sqrt{3}i}{2} = \cos\frac{\pi}{3} - i\sin\frac{\pi}{3}.$$

This equation means (geometrically) that a_3 is the image of a_1 under rotation through $-60°$ about a_2. Thus the triangle $A_1A_2A_3$ is equilateral.

1986/3

First Solution. Note that the sum of the five numbers is an invariant of the process; its value remains unchanged in each step. Denote this fixed value of the sum by S. By hypothesis, $S > 0$.

The idea is to find a monovariant, i.e., a quantity whose value keeps decreasing during the procedure. If the changes are strictly monotonic and the values are positive integers, the process must terminate. This job can be done, for example, by the function f defined as follows: if x_1, x_2, x_3, x_4, x_5 are the five numbers assigned to the vertices of the pentagon, set

$$f(x_1, x_2, x_3, x_4, x_5) = \frac{1}{2}\sum_{i=1}^{5}(x_{i+1} - x_{i-1})^2 = \sum_{i=1}^{5} x_i^2 - \sum_{i=1}^{5} x_{i-1}x_{i+1};$$

in cyclic labeling, $x_0 = x_5$, $x_6 = x_1$.

Assume that at a certain moment at least one of the five numbers is negative; labeling can be fixed so that, e.g., $x_3 < 0$, and the operation defined in the problem statement is performed on choosing $x = x_2$, $y = x_3$, $z = x_4$. Write for convenience $u = x_1$, $w = x_5$. This operation takes the vector $(x_1, x_2, x_3, x_4, x_5) = (u, x, y, z, w)$ into

$$(x_1', x_2', x_3', x_4', x_5') = (u, x + y, -y, y + z, w).$$

Let us examine the change of the value of f:

$$\begin{aligned}\sum_{i=1}^{5}(x_i')^2 - \sum_{i=1}^{5} x_i^2 &= (x + y)^2 + (y + z)^2 - x^2 - z^2\\ &= 2xy + 2y^2 + 2yz,\end{aligned}$$

$$\begin{aligned}&\sum_{i=1}^{5} x_{i-1}x_{i+1} - \sum_{i=1}^{5} x_{i-1}'x_{i+1}'\\ &= (wx + uy + xz + yw + zu)\\ &\quad - (w(x + y) - uy + (x + y)(y + z) - yw + (y + z)u)\\ &= uy - xy - y^2 - yz + yw.\end{aligned}$$

Adding these two equalities we obtain

$$\begin{aligned}f(x_1', x_2', x_3', x_4', x_5') - f(x_1, x_2, x_3, x_4, x_5) &= uy + xy + y^2 + yz + yw\\ &= Sy < 0.\end{aligned}$$

Thus, indeed, the values of f form a strictly decreasing sequence of positive integers and the process must come to an end.

Second Solution. (An outline.) The idea is the same as the first solution: to construct a certain monovariant. Now we assign to the vector $(x_1, x_2, x_3, x_4, x_5)$ the number $g(x_1, x_2, x_3, x_4, x_5) = A + B + C + D$ where (in cyclic labeling)

$$A = \sum_{i=1}^{5} |x_i|, \qquad B = \sum_{i=1}^{5} |x_i + x_{i+1}|,$$

$$C = \sum_{i=1}^{5} |x_i + x_{i+1} + x_{i+2}|, \qquad D = \sum_{i=1}^{5} |x_i + x_{i+1} + x_{i+2} + x_{i+3}|.$$

Each one of the expressions A, B, C, D is a sum of five addends. Thus the number $g(x_1, x_2, x_3, x_4, x_5)$ is a sum of twenty addends.

Take a vector $(x_1, x_2, x_3, x_4, x_5)$ with at least one $x_k < 0$ and transform it into $(x_1', x_2', x_3', x_4', x_5')$, as in the first solution. Upon comparing $G = g(x_1, x_2, x_3, x_4, x_5)$ with $G' = g(x_1', x_2', x_3', x_4', x_5')$, one can verify that out of twenty summands that make up the number G', nineteen summands are present in the representation of G. Thus, in calculating the difference $G' - G$ we have an almost complete obliteration, and it can be checked that the non-obliterating summands yield the outcome

$$G' - G = |S + x_k| - |S - x_k|.$$

Since $x_k < 0$ and $S > 0$, the number $|S - x_k|$ exceeds $|S + x_k|$ in absolute value, showing that $G' - G < 0$; the result follows.

Remark. The second solution, by Joseph Keane, the IMO contestant (from the USA team), was awarded a special prize.

1986/4

Let C be the vertex adjacent to B, other than A, and assume that Y and Z are interior points of the sides AB and BC respectively. The condition $n \geq 5$ ensures that, in each moment, Y and Z lie on adjacent sides of the polygon. The circumcircle of triangle XYZ passes through B because $\angle YXZ + \angle YBZ = \angle AOB + \angle ABO + \angle OBC = 180°$. The angles XYZ and XBZ, subtended by arc XZ of that circle, are equal. Hence $\angle OBZ = \angle OBC = \angle XYZ = \angle XBZ$, showing that the points B, O, X are collinear. X cannot lie between O and B (in which case its distance to Y or Z would

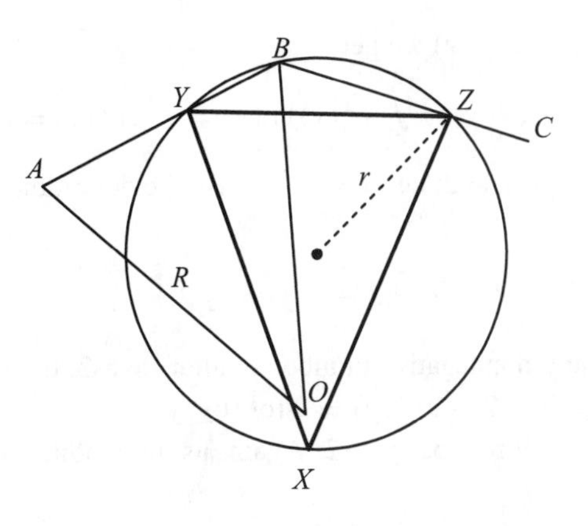

be smaller than OB, in contradiction to OAB and XYZ being congruent triangles). Thus X lies on line BO beyond O.

Denote by R the circumradius of the n-gon and by r the circumradius of triangle OAB. Then $R = 2r \sin \angle OAB = 2r \cos(180^\circ/n)$. The circumcircle of $BYXZ$ (i.e., of triangle XYZ) has also radius r. Its chord BX is the longest possible when it is a diameter of the circle. Hence the maximum length of the segment OX, attained when the points Y and Z are symmetric across line OB, is

$$d = 2r - R = R\left(\frac{1}{\cos(180^\circ/n)} - 1\right).$$

When the points Y and Z move along the sides AB and BC, the point X remains on the line segment of length d producing BO; by continuity, each point of that segment is an available position of X. Accordingly, as Y and Z trace out the whole boundary of the polygon, the locus of X is the asterisk composed of n line segments of length d, emanating from O and pointing away from the vertices.

1986/5

Let f be a function satisfying the given conditions. Fix y with $0 \le y < 2$. Then $f(y) \ne 0$ by (iii), whereas by (ii) and (i) we get

$$0 = f(2) = f((2-y)+y) = f((2-y)f(y))f(y).$$

So $f((2-y)f(y)) = 0$, which in view of (iii) implies the inequality

$$(2-y)f(y) \ge 2.$$

Setting in (i) $x = 2/f(y)$ we get

$$f(x+y) = f(xf(y))f(y) = f(2)f(y) = 0.$$

Hence by (iii) $x + y \geq 2$; i.e., $x \geq 2 - y$. By the definition of x, this means that

$$(2-y)f(y) \leq 2.$$

Since y was any nonnegative number smaller than 2, the two inequalities obtained show that $(2-y)f(y) = 2$ for $0 \leq y < 2$.

Now take any number $y \geq 2$. Again, using conditions (i) and (ii) we see that

$$f(y) = f((y-2)+2) = f((y-2)f(2))f(2) = 0.$$

The result follows: f must be the function

$$f(y) = \begin{cases} \dfrac{2}{2-y} & \text{for} \quad 0 \leq y < 2, \\ 0 & \text{for} \quad y \geq 2. \end{cases}$$

It is a routine task to verify that this function indeed satisfies all the conditions of the problem.

1986/6

First Solution. The answer is yes; it is always possible to color the points as required. For a proof, take an arbitrary line L parallel to one of the axes and intersecting the given set A. Let $P_1, P_2, \ldots, P_r$ be the points of A on line L, arranged in the order of increasing free coordinate. We connect P_1 with P_2, P_3 with P_4, and so on, by line segments; there may remain a single odd point.

The same is done on every such line L, horizontal or vertical. We obtain a family of segments, each point of A belonging to two segments, at most; if two segments have a common endpoint, they are perpendicular.

The union of these segments splits into polygonal lines, closed or not, without common vertices. Those which are closed are composed of an even number of edges because adjacent segments are perpendicular.

We color the vertices of each polygonal line alternately: red, white, red, white, etc.; this is possible, in view of the observation on the evenness of cycles. If there have remained any loose points in A, not belonging to any of the segments drawn, we color them arbitrarily. The coloring thus obtained fulfils the requirement of the problem; points on each horizontal or vertical line are connected pairwise by segments with endpoints of different colors; if there are an odd number of points on a line, the color of the remaining (rightmost or uppermost) point is insignificant.

Second Solution. Such a coloring always exists; we prove this by induction on the number of points in the given set A; when A is a singleton, there is nothing to prove. Fix $n \geq 2$ and assume that every set of cardinality less than n can be colored in the desired manner. Let A be an n-point set. Consider two cases.

Case 1. There exists a horizontal or vertical line L with an odd number of points of the set A on it. Choose one of these points, call it P, and remove it from A. Color the points of the remaining $(n-1)$-element set red and white, according to the induction hypothesis. An even number of points remain on line L, half of them red, the other half white.

Let L' be the line passing through P and perpendicular to L. The number of red points (from the set $A \setminus \{P\}$) on L' differs from the number of white points by at most 1. So we can choose a color for the point P without violating this property. On line L we now have one red point more or less than there are white points.

Case 2. Every line, horizontal or vertical, contains an even number of points of the set A. Let P be any point of A and let L and L' be the two perpendicular lines through P. As before, we remove point P, obtaining an $(n-1)$-element set, which by induction hypothesis can be colored as required. On each horizontal or vertical line other than L and L' we have perfect balance.

On line L there is an odd number of points of the set $A \setminus \{P\}$; the numbers of red points and white points differ by 1. Without loss of generality assume there is one white point more. This means that in the whole set $A \setminus \{P\}$ the number of white points exceeds by 1 the number of red points. The same can be said about the numbers of white and red points on line L' (because these numbers are balanced on every other line parallel to L'). If we now color the point P red, we achieve balance on both lines L and L'.

In each case we have found a coloring of the points of set A as needed. It follows by induction that a good coloring exists for every set A, of arbitrary cardinality.

Twenty-eighth International Olympiad, 1987

1987/1

First Solution. There are $p_n(k)$ permutations of the set $S = \{1, \ldots, n\}$ which have exactly k fixed points. Therefore

$$F(n) = \sum_{k=0}^{n} k \cdot p_n(k) \tag{1}$$

is the number of all fixed points in all permutations. Each element $j \in S$ is a fixed point of $(n-1)!$ permutations—namely, those of the form

$$f(i) = \begin{cases} j & \text{for } \ i = j, \\ g(i) & \text{for } \ i \neq j, \end{cases}$$

where g is any permutation of the $(n-1)$-element set $S \setminus \{j\}$.

As there are n ways to select an element $j \in S$ and $(n-1)!$ ways to choose a permutation g of the set $S \setminus \{j\}$, the total value of $F(n)$ is $n \cdot (n-1)!$; in short, $F(n) = n!$.

Second Solution. Obviously, $p_n(k) = 0$ for $k < 0$ and for $k > n$. The number of all permutations of the set $\{1, \ldots, n\}$, with arbitrarily many fixed points, is expressed by

$$\sum_{k=0}^{n} p_n(k) = n!. \tag{2}$$

We will prove the following recurrence relation (for $n \geq 3$):

$$p_n(k) = p_{n-1}(k-1) + (n-1)(p_{n-2}(k) + p_{n-1}(k) - p_{n-2}(k-1)). \tag{3}$$

When $k < 0$ or $k > n$, both sides of (3) are zero. For the sequel assume $0 \leq k \leq n$.

Let $S = \{1, \ldots, n\}$, $S' = \{1, \ldots, n-1\}$, and let f be a permutation of S with exactly k fixed points. Consider three cases:

(a) the element n is a fixed point of f;

(b) the element n belongs to a cycle of length 2;

(c) the element n belongs to a cycle of length greater than 2.

In case (a), f restricted to S' is a permutation of S' with exactly $k-1$ fixed points. There are $p_{n-1}(k-1)$ such permutations.

In case (b), there exists an $m \in S'$ such that $f\colon m \mapsto n \mapsto m$; then f, restricted to the set $S \setminus \{m, n\}$, is a permutation of this set with exactly k

fixed points. Since we have $n-1$ possibilities of choosing the element m, there are $(n-1)p_{n-2}(k)$ such permutations.

In case (c), there exists an $m \in S'$ such that $f\colon m \mapsto n \mapsto r$, where $r \in S'$, $r \neq m$. We associate with f the permutation g of the set S' which sends m directly to r (skipping n), and otherwise coincides with f. Then g has k fixed points, but the element m is *not* one of them. For a specific m, there are $p_{n-1}(k) - p_{n-2}(k-1)$ such permutations (the subtracted term $p_{n-2}(k-1)$ expresses the number of those permutations of S' that have k fixed points, m being one of them). Taking into account the $n-1$ possible choices of m, we get the number of $(n-1)(p_{n-1}(k) - p_{n-2}(k-1))$ such permutations g. And conversely, every such permutation g corresponds to a unique f. (These considerations are correct also in the case when some terms are of the form $p_\ell(j)$ with $j < 0$ or $j > \ell$, and hence are equal to zero.)

Adding up the outcomes in cases (a), (b), (c), we obtain the recurrence formula (3). It is then a matter of simple calculation to show (using equations (2) and (3)) that the number $F(n)$ defined by (1) satisfies for $n \geq 3$ the recursion $F(n) = nF(n-1)$. And since $p_1(1) = 1$, $p_2(1) = 0$, $p_2(2) = 1$, we get $F(1) = 1$, $F(2) = 2$, and by induction $F(n) = n!$.

Remark. As compared to the first solution, the second one is much longer and less elegant; but it gives more information about the particular summands of the sum in question; i.e., the numbers $p_n(k)$. For instance, taking $k = 0$ we obtain the formula $p_n(0) = (n-1)(p_{n-2}(0) + p_{n-1}(0))$ which allows us to compute the number of fixed-point free permutations.

The assertion of the problem admits a graceful interpretation: an absent-minded person wrote n letters, prepared n envelopes with addresses, and put the letters in envelopes hastily, in a random way. The probability that exactly k letters are put into the proper envelopes equals $p_n(k)/n!$. The number $F(n)/n!$ is the expected value of letters placed correctly; it is equal to 1, according to the result of the problem. This means that, on the average, just one letter (independently of n) will reach the right addressee.

1987/2

First Solution. Points K and M are situated symmetrically with respect to line AL, the bisector of angle A. Triangles AKN and AMN are therefore congruent, and so

$$S(AKNM) = 2S(AKN) = AK \cdot AN \cdot \sin\varphi,$$

where $\alpha = 2\varphi = \angle CAB$ and the symbol $S(XY\ldots)$ denotes the area of the polygonal figure $XY\ldots$. Since $AK = AL \cdot \cos\varphi$, we get

$$S(AKNM) = AL \cdot \cos\varphi \cdot AN \cdot \sin\varphi = \tfrac{1}{2}AL \cdot AN \cdot \sin\alpha.$$

Angles BNA and BCA, subtended by arc AB of the circumcircle of triangle ABC, are equal. Angles NAB and CAL are also equal (by hypothesis). It follows that triangles ABN and ALC are similar, and therefore $AB/AL = AN/AC$. Thus, finally,

$$S(AKNM) = \tfrac{1}{2}AL \cdot AN \cdot \sin\alpha = \tfrac{1}{2}AB \cdot AC \cdot \sin\alpha = S(ABC).$$

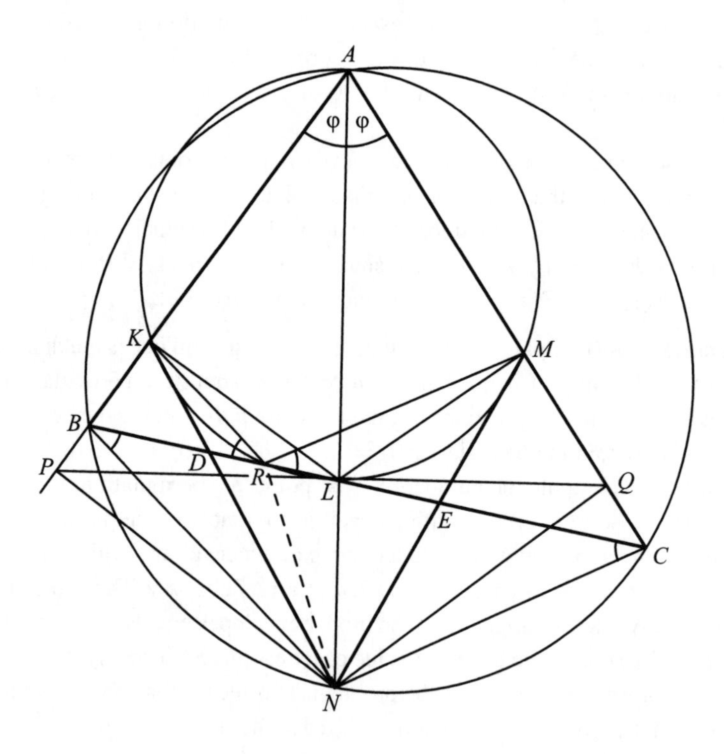

Second Solution. Let P and Q be the feet of perpendiculars drawn from N to lines AB and AC. Assume $AB \le AC$ without loss of generality; then P lies on ray AB beyond B, while Q lies on segment MC. Since AN bisects angle CAB, we have $KL = ML$ and $PN = QN$. Equal angles BAN and CAN are subtended by equal arcs BN and CN of the circumcircle. So the chords BN and CN are equal, and consequently the right triangles BPN and CQN are congruent. Hence $BP = CQ$.

We now express the areas in question as follows:

$$\begin{aligned} S(ABC) &= S(AKLM) + S(KBL) + S(MCL); \\ S(AKNM) &= S(AKLM) + S(KNL) + S(MNL) \\ &= S(AKLM) + S(KPL) + S(MQL); \end{aligned}$$

the last equality holds because $KL \parallel PN$ and $ML \parallel QN$. Subtraction yields

$$\begin{aligned} S(ABC) - S(AKNM) &= S(KBL) + S(MCL) - S(KPL) - S(MQL) \\ &= \tfrac{1}{2}KB \cdot KL + \tfrac{1}{2}MC \cdot ML \\ &\quad - \tfrac{1}{2}KP \cdot KL - \tfrac{1}{2}MQ \cdot ML \\ &= \tfrac{1}{2}ML \cdot (MC - MQ) - \tfrac{1}{2}KL \cdot (KP - KB) \\ &= \tfrac{1}{2}ML \cdot CQ - \tfrac{1}{2}KL \cdot BP. \end{aligned}$$

And since we have shown that $KL = ML$ and $BP = CQ$, the difference in the last line is equal to zero.

Third Solution. First assume $AB \neq AC$; let e.g., $AB < AC$. Then the circle with diameter AL intersects line BC in two points, L and R. Points B and C lie outside this circle, since ABC is an acute triangle. So R lies on segment BC, between B and L. Points K and M (the orthogonal projections of L on lines AB and AC) lie on that circle.

Let as before $\varphi = \frac{1}{2}\angle CAB$. The angles in the following equalities are either subtended by equal arcs of the circles under consideration or on arcs mutually complementing to the whole circle:

$$\begin{gathered} \angle MRC = \angle MRL = \angle MAL = \varphi, \\ \angle KRB = 180^\circ - \angle KRL = \angle KAL = \varphi, \\ \angle BCN = \angle BAN = \varphi, \quad \angle CBN = \angle CAN = \varphi. \end{gathered}$$

From these equalities ($\angle KRB = \angle CBN$ and $\angle MRC = \angle BCN$) it follows that $BN \parallel KR$ and $CN \parallel MR$. This is of course true also in the case where $AB = AC$ (R coinciding with L).

Let line BC intersect segments KN and MN in the respective points D and E. Then

$$\frac{S(KBD)}{S(KDR)} = \frac{BD}{DR} = \frac{ND}{KD} = \frac{S(NRD)}{S(KDR)};$$

the middle equality follows from $BN \parallel KR$. Thus

$$S(KBD) = S(NRD); \quad \text{analogously,} \quad S(MCE) = S(NRE).$$

And since

$$S(ABC) = S(AKDEM) + S(KBD) + S(MCE),$$
$$S(AKNM) = S(AKDEM) + S(NRD) + S(NRE),$$

the claimed equality $S(ABC) = S(AKNM)$ results.

1987/3

Assume (relabeling if necessary) that the negative x_is, if any, are the final terms of the sequence; thus there is an m such that $x_1, \ldots, x_m \geq 0$ and $x_{m+1}, \ldots, x_n < 0$ (if all x_is are negative, then $m = 0$; if all x_is are nonnegative, then $m = n$).

Define $\mathcal{C}$ to be the set of all n-tuples $(c_1, c_2, \ldots, c_n)$ with integer terms $c_i \in \{0, 1, \ldots, k-1\}$. Consider all the values that the sum

$$S = c_1x_1 + c_2x_2 + \cdots + c_nx_n$$

can take for $(c_1, c_2, \ldots, c_n)$ from $\mathcal{C}$. The least value and the greatest value of S are attained respectively for

$$(c_1, c_2, \ldots, c_n) = (\underbrace{0, \ldots, 0}_{m}, \underbrace{k-1, \ldots, k-1}_{n-m}),$$
$$(c_1, c_2, \ldots, c_n) = (\underbrace{k-1, \ldots, k-1}_{m}, \underbrace{0, \ldots, 0}_{n-m}).$$

Denote these extreme values by A and B (respectively); thus

$$B - A = (k-1)(x_1 + \cdots + x_m) - (k-1)(x_{m+1} + \cdots + x_n)$$
$$= (k-1)(|x_1| + |x_2| + \cdots + |x_n|).$$

Since by hypothesis $x_1^2 + x_2^2 + \cdots + x_n^2 = 1$, we obtain

$$|x_1| + |x_2| + \cdots + |x_n| \leq \sqrt{n}$$

by Cauchy–Schwarz inequality (see *Glossary*). So $B - A \leq (k-1)\sqrt{n}$.

All available values of S lie in the interval $[A, B]$. Partition it into $N = k^n - 1$ equal subintervals, each of length $(B - A)/N$. The set $\mathcal{C}$ contains k^n tuples; so we can find two distinct n-tuples $(c_1', c_2', \ldots, c_n')$ and $(c_1'', c_2'', \ldots, c_n'')$ producing values of the sum S that belong to the same subinterval. Denote these values by S' and S'':

$$S' = c_1'x_1 + \cdots + c_n'x_n, \quad S'' = c_1''x_1 + \cdots + c_n''x_n; \quad |S' - S''| \leq \frac{B - A}{N}.$$

Taking $a_i = c_i' - c_i''$ for $i = 1, 2, \ldots, n$ we obtain a sequence of integers $(a_1, a_2, \ldots, a_n) \neq (0, 0, \ldots, 0)$ satisfying the inequalities $|a_i| \leq k - 1$ for $i = 1, 2, \ldots, n$ and

$$|a_1x_1 + a_2x_2 + \cdots + a_nx_n| = |S' - S''| \leq \frac{B - A}{N} \leq \frac{(k-1)\sqrt{n}}{k^n - 1};$$

so we have found integers $a_1, a_2, \ldots, a_n$ with properties as needed.

1987/4

First Solution. Replace 1987 by a general positive integer c. Let N_0 be the set of all nonnegative integers. We are going to find out for what values of c there exists a function $f : N_0 \to N_0$ satisfying the equation

$$f(f(n)) = n + c \quad \text{for } n \in N_0. \tag{1}$$

Fix an integer $c \geq 1$ and assume f is such a function. Then

$$f(r + c) = f(f(f(r))) = f(r) + c \quad \text{for } r \in N_0, \tag{2}$$

and by induction

$$f(r + cq) = f(r) + cq \quad \text{for } q, r \in N_0. \tag{3}$$

Let $C = \{0, 1, \ldots, c - 1\}$. For any $n \in C$ let $g(n)$ and $h(n)$ be the quotient and remainder left by $f(n)$ in division by c:

$$f(n) = h(n) + cg(n); \qquad g(n) \in N_0, \quad h(n) \in C. \tag{4}$$

Applying equation (3) to $q = g(n)$, $r = h(n)$ we get

$$f(f(n)) = f(h(n)) + cg(n) \geq cg(n).$$

Since $n \in C$, equation (1) yields $f(f(n)) = n + c < 2c$. So $cg(n) < 2c$, i.e., $g(n) = 0$ or $g(n) = 1$. (Hence by (4), $f(n) < 2c$ for $n \in C$; this inequality will be needed in the second solution.)

Consider the sets

$$C_0 = \{n \in C\colon\ g(n) = 0\}, \qquad C_1 = \{n \in C\colon\ g(n) = 1\}.$$

They are disjoint, and their union is the whole of C.

If $n \in C_0$, then by (4) $f(n) = h(n)$, and according to (1)

$$f(h(n)) = f(f(n)) = n + c.$$

Dividing $f(h(n))$ by c we thus get quotient 1 and remainder n:

$$g(h(n)) = 1, \qquad h(h(n)) = n \quad \text{for } n \in C_0.$$

And if $n \in C_1$, then by (4) $f(n) = h(n) + c$. Using equations (1) and (2) (for $r = h(n)$) we obtain

$$n + c = f(f(n)) = f(h(n) + c) = f(h(n)) + c.$$

Thus $f(h(n)) = n$, which means that

$$g(h(n)) = 0, \qquad h(h(n)) = n \quad \text{for } n \in C_1.$$

Properties of the function $h\colon C \to C$ can be now summarized as follows:

$$h(h(n)) = n \quad \text{for } n \in C; \tag{5}$$

$$n \in C_0 \Longrightarrow h(n) \in C_1; \qquad n \in C_1 \Longrightarrow h(n) \in C_0. \tag{6}$$

Equality (5) shows that h is a one-to-one mapping of the set C onto itself. Hence and from (6) we infer that h maps the set C_0 onto (the whole of) C_1, and the set C_1 onto C_0. So these sets contain equally many elements, and consequently the set $C = C_0 \cup C_1$ has an even number of elements.

Recall that C is a c-element set. The conclusion follows: a function $f\colon \mathrm{N}_0 \to \mathrm{N}_0$ satisfying equation (1) can exist only if c is an even integer; there is no such function for $c = 1987$.

(And conversely, it is clear that if c is even, then f does exist; it suffices to define $f(n) = n + \frac{1}{2}c$.)

Second Solution. As in the first solution, we fix an integer $c \geq 1$, assume that $f\colon \mathrm{N}_0 \to \mathrm{N}_0$ is a function satisfying equation (1) and derive formula (3). Defining the set $C = \{0, 1, \ldots, c-1\}$ we arrive at the conclusion that $f(r) < 2c$ for $r \in C$.

Let k be any positive integer. Every number $n \in \mathrm{N}_0$ smaller than kc can be written uniquely in the form $n = r + cq$ with $r \in C$, $q \in \mathrm{N}_0$, $q \leq k-1$. Since $f(r) < 2c$, formula (3) yields for $n = r + cq$ the inequality

$$f(n) < 2c + cq \leq 2c + (k-1)c = (k+1)c.$$

We have thus shown that for every $n \in \mathrm{N}_0$ and for $k = 1, 2, 3, \ldots$:

$$\text{if} \quad n < kc, \qquad \text{then} \quad f(n) < (k+1)c. \tag{7}$$

Consider the sets

$$A = \{n \in \mathrm{N}_0\colon\ f(n) - n \text{ is even}\}, \qquad B = \{n \in \mathrm{N}_0\colon\ f(n) - n \text{ is odd}\}$$

and their subsets

$$A_k = \{n \in A\colon\ n < kc\}, \qquad B_k = \{n \in B\colon\ n < kc\}$$

for $k = 1, 2, 3, \ldots$. From formula (2) of the first solution we obtain the equality $f(n+c) - (n+c) = f(n) - n$, which shows that if $n \in A$, then $n + c \in A$, and if $n \in B$, then $n + c \in B$.

Suppose that the interval $[0, c)$ contains a elements of the set A and b elements of B; then $a + b = c$. It follows that in each interval $[kc, (k+1)c)$ there are a elements of A and b elements of B. Thus, for each k, the set A_k has ak elements, and the set B_k has bk elements.

We are going to prove that c must be an even number. Assume to the contrary that c is odd. Take an $n \in \mathrm{N}_0$ and write $m = f(n)$. From equation (1) we get $[f(m) - m] + [f(n) - n] = f(f(n)) - n = c$. This shows that if $n \in A$, then $m \in B$ (i.e., $f(n) \in B$), and vice versa. Hence, in view of (7), f maps the set A_k into B_{k+1} and the set B_k into A_{k+1}.

From (1) it is clear that f is injective. This implies the inequalities $ka \le (k+1)b$ and $kb \le (k+1)a$; equivalently:

$$\frac{k}{k+1} \le \frac{a}{b} \le \frac{k+1}{k} \qquad \text{for } k = 1, 2, 3, \ldots.$$

And since k can be arbitrarily large, we eventually conclude $a = b$. This, however, contradicts the assumption that $c = a + b$ is odd.

In conclusion, a function satisfying equation (1) can exist only if c is even; in particular, it does not exist for $c = 1987$.

1987/5

The idea is to consider lattice points on some simple algebraic curve (in a fixed Cartesian coordinate system on the plane). A triangle with vertices in lattice points (i.e., points with integer coordinates) will be called a lattice triangle.

The area of a lattice triangle is either an integer or half an integer. This follows from the fact that every such triangle can be obtained from a rectangle with sides of integer lengths by cutting off a few right triangles with legs of integer lengths.

On the parabola $y = x^2$ mark the lattice points $P_k = (k, k^2)$ (for $k = 1, \ldots, n$); no three of them are collinear. In other words, any three of them span a nondegenerated triangle. As noticed, the areas of all these lattice triangles are rational. If $k \neq m$, then

$$P_k P_m = \sqrt{(k-m)^2 + (k^2 - m^2)^2} = |k - m|\sqrt{1 + (k+m)^2}.$$

The square root of an integer is a rational number if and only if it is an integer. And since $k + m > 0$, the number $(k+m)^2 + 1$ is not a perfect square. Thus all the distances $P_k P_m$ are irrational, as desired.

Remark. There are many other examples of n-element sets of lattice points, no three in line, with all the mutual distances irrational; we invite the reader to find some.

1987/6

Write $f(k) = k^2 + k + n$. Let m be the least nonnegative integer for which $f(m)$ is a composite number. By hypothesis, $m > \sqrt{n/3}$; hence $n < 3m^2$. Our task is to prove that $m \geq n - 1$.

Let p be the smallest prime divisor of the (composite) number $f(m)$. Obviously, $p \leq \sqrt{f(m)}$; hence

$$p^2 \leq f(m) = m^2 + m + n < m^2 + m + 3m^2 < (2m+1)^2,$$

and consequently $p \leq 2m$.

Note the identity $f(m) - f(k) = (m-k)(m+k+1)$. Setting for k the integers from 0 to $m-1$ and multiplying out the equalities that result, we obtain

$$\prod_{k=0}^{m-1} (f(m) - f(k)) = \prod_{k=0}^{m-1} (m-k)(m+k+1) = (2m)!$$

The number $(2m)!$ is divisible by p (as $p \leq 2m$). Consequently there exists an $\ell \in \{0, 1, \ldots, m-1\}$ such that $(m-\ell)(m+\ell+1)$ is divisible by p. Hence $p \leq m + \ell + 1$.

The prime p divides the number $f(m)$ and the difference $f(m) - f(\ell)$; hence it is also a divisor of $f(\ell)$. But $f(\ell)$ is a prime, according to the definition of m. Therefore $f(\ell) = p$; equivalently: $p - \ell = \ell^2 + n$.

This combined with the former inequality $p - \ell \leq m + 1$ yields the claimed result: $m + 1 \geq \ell^2 + n \geq n$.

Twenty-ninth International Olympiad, 1988

1988/1

(i) The line symmetric to BC with respect to O (the common center of the two circles) passes through A and intersects the larger circle at D and E, so that the points B and D lie on one side of line ℓ, and C and E lie on the other side (the reasoning that follows is case-independent).

Denote the lengths of PB and PC by u and v, respectively. The well-known equality $uv = R^2 - r^2$ holds for any position of point B (see *Glos-*

sary: *Power of a point*). Quadrilaterals $BPAD$ and $BCED$ are rectangles (degenerating to segments when BC is a diameter). Therefore

$$PA^2 = BD^2 = CD^2 - BC^2 = 4R^2 - (u+v)^2;$$

and hence

$$\begin{aligned} BC^2 + CA^2 + AB^2 &= BC^2 + PC^2 + PA^2 + PA^2 + PB^2 \\ &= (u+v)^2 + v^2 + 2\big(4R^2 - (u+v)^2\big) + u^2 \\ &= 8R^2 - 2uv = 6R^2 + 2r^2. \end{aligned}$$

Thus the set of values of the sum in question consists of the single number $6R^2 + 2r^2$.

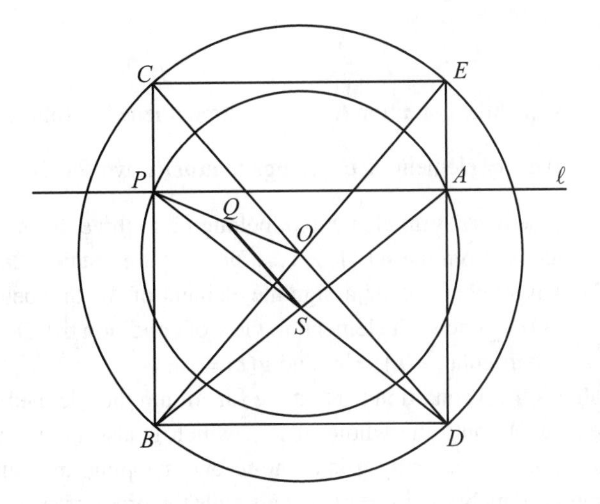

(ii) Let S be the common midpoint of the diagonals AB and PD of rectangle $BPAD$ and let Q be the midpoint of OP. Segment QS, connecting the midpoints of two sides of triangle OPD, has length $R/2$. Thus S lies on the circle with center Q and radius $R/2$.

To show that every point of that circle is an admissible position of the midpoint of AB, we reverse the reasoning: let S be any point such that $QS = R/2$. Ray PS cuts the big circle at a point which we call D. Ray DO cuts the big circle at a point C. Ray CP cuts the big circle at a point B. Finally, ray BO cuts the big circle at a point E. We obtain a rectangle $BCED$, symmetric about O.

One of the intersection points of line DE with the small circle—call it A—completes the rectangle $PBDA$ (we invite the reader to make a diagram; points O, P, hence also Q, are given; points S, D, C, B, E, A should be marked on the sketch in this order, in accordance with the construction above).

Since $OP < R$, we get $QP < R/2$, which means that P lies inside the circle of radius $R/2$ centered at Q. Consequently, segment PD has no point in common with this circle other than S. In other words, S is the only point of side PD of triangle OPD at distance $R/2$ from Q. Thus S must be the midpoint of side PD. And since $BPAD$ is a rectangle, S must be the midpoint of AB as well, and this is exactly what had to be proved.

In conclusion, the locus of the midpoint of AB is the circle of radius $R/2$, centered at the midpoint of OP.

1988/2

We first show that the condition (c) can be strengthened as follows:

(d) every element of B belongs to *exactly* two sets A_i.

Indeed: suppose there is an element x_0 belonging to three distinct sets A_k, A_ℓ, A_m. Delete m from the set $\{1, 2, \ldots, 2n + 1\}$; we obtain a $2n$-element set M. To each number $j \in M$ assign the element of A_m that belongs also to A_j; there is only one such element, in view of condition (b). Denote it by $g(j)$; then, in particular, $g(k) = x_0$ and $g(\ell) = x_0$.

It follows from condition (c) that the function g thus defined maps the $2n$-element set M onto the whole of A_m, which is also a $2n$-element set (condition (a)). Consequently g is a one-to-one mapping, in contradiction to the previous equality $g(k) = g(\ell)$. This ends the proof of property (d).

On account of conditions (b) and (d), the elements of B can be identified with unordered pairs (i.e., two-element sets) $\{i, j\}$ with entries from the set $\{1, 2, \ldots, 2n + 1\}$; to make things clear: an element $x \in B$ is identified with the unique pair $\{i, j\}$ $(i \neq j)$ such that $A_i \cap A_j = \{x\}$.

Now consider a convex (e.g., regular) polygon P with $2n + 1$ vertices labeled $1, 2, \ldots, 2n+1$. The vertex given label i will represent the set A_i. The elements of B, i.e., pairs $\{i, j\}$ $(i \neq j)$, will be represented by lines (sides or diagonals) connecting the respective vertices; the line between vertices i and j represents the unique element of $A_i \cap A_j$. (Readers familiar with the language of graph theory will clearly recognize a complete graph in this description.)

Instead of the symbols 0 and 1, we will rather speak about two colors; say, grey and pink. The problem can be now translated as follows: for what values of n is it possible to color the sides and diagonals of P in these two colors in such a way that each vertex should issue n lines of either color?

If this is possible, then there are equally many grey lines and pink lines. The joint number of all lines, equal to $n(2n+1)$, has to be even, which means that n must be even.

And conversely, suppose n is even. Then the coloring with the properties as needed can be done for instance in the following fashion. Color pink all the sides of P; color grey each diagonal skipping a single vertex; and so on: if two vertices are separated by k sides ($k \le n$), then the diagonal connecting these two vertices has to be colored grey if k is even and pink if k is odd. Since n is even, we see that every vertex will be an endpoint of n grey lines and n pink lines.

The answer follows: one can assign to every element of B (i.e., to every pair $\{i, j\}$, $i \ne j$) one of two given symbols, observing the postulated condition, if and only if n is an even integer.

Remark. In the original terminology, the assignment proposed above (for an even n) corresponds to defining a function $f: B \to \{0, 1\}$ by the formula

$$f(x) = \begin{cases} 1 & \text{if } x \in A_i \cap A_j, \\ & j - i \in \{2, 4, \ldots, n\} \cup \{n+1, n+3, \ldots, 2n-1\}, \\ 0 & \text{if } x \in A_i \cap A_j, \\ & j - i \in \{1, 3, \ldots, n-1\} \cup \{n+2, n+4, \ldots, 2n\}. \end{cases}$$

1988/3

The given equations determine the function f uniquely because every integer N greater than 1 and different from 3 can be written in the form $N = 2n$ or $N = 4n + 1$ or $N = 4n + 3$, for a certain positive integer n, less than N. Calculating the values of $f(n)$ for small integers n we soon arrive at the guess that f reverses the binary representation of n; e.g., for $n = 19 = (10011)_2$ we get $f(19) = 25 = (11001)_2$. In other words: we presume that f coincides with the function g defined by the rule:

$$\text{if} \quad n = (c_m c_{m-1} c_{m-2} \ldots c_2 c_1 c_0)_2 \quad \text{and} \quad c_m = 1,$$
$$\text{then} \quad g(n) = (c_0 c_1 c_2 \ldots c_{m-2} c_{m-1} c_m)_2;$$

we do not exclude leading zeros in the representation of $g(n)$.

Note that $g(1) = 1$, $g(3) = 3$. To prove that $f(n) = g(n)$ for every n, it will be enough to verify that the function g obeys the same recursion relations as those imposed on f in the problem statement. Consider an arbitrary positive integer $n = (c_m c_{m-1} \ldots c_1 c_0)_2$ with the leading digit $c_m = 1$. Then

$$2n = (c_m c_{m-1} \ldots c_1 c_0 0)_2, \quad 2n+1 = (c_m c_{m-1} \ldots c_1 c_0 1)_2,$$
$$4n+1 = (c_m c_{m-1} \ldots c_1 c_0 01)_2, \quad 4n+3 = (c_m c_{m-1} \ldots c_1 c_0 11)_2.$$

According to the definition of g,

$$g(2n) = (c_0 c_1 \ldots c_{m-1} c_m)_2 = g(n),$$
$$g(2n+1) = (1 c_0 c_1 \ldots c_{m-1} c_m)_2 = (1\underbrace{00\ldots00}_{m+1})_2 + (c_0 c_1 \ldots c_m)_2,$$
$$g(4n+1) = (10 c_0 c_1 \ldots c_{m-1} c_m)_2 = (10\underbrace{00\ldots00}_{m+1})_2 + (c_0 c_1 \ldots c_m)_2,$$
$$g(4n+3) = (11 c_0 c_1 \ldots c_{m-1} c_m)_2 = (11\underbrace{00\ldots00}_{m+1})_2 + (c_0 c_1 \ldots c_m)_2;$$

equivalently,

$$g(2n) = g(n), \qquad g(2n+1) = 2^{m+1} + g(n),$$
$$g(4n+1) = 2^{m+2} + g(n), \quad g(4n+3) = 2^{m+2} + 2^{m+1} + g(n).$$

The recursion relations follow: $g(2n) = g(n)$ and

$$g(4n+1) = 2g(2n+1) - g(n), \quad g(4n+3) = 3g(2n+1) - 2g(n).$$

They are identical with those that define the function f. This means that f is the same function as g; our guess was correct.

A positive integer $n = (c_m c_{m-1} \ldots c_1 c_0)_2$ with the leading $c_m = 1$ satisfies the condition $f(n) = n$ if and only if its binary expansion is symmetric; i.e., when $c_k = c_{m-k}$ for $k = 0, \ldots, m$; numbers with this property are often called *palindromes*.

Fix $m \geq 0$. All the integers from the interval $[2^m, 2^{m+1})$ have $(m+1)$-digit binary expansions with a leading *one*. If m is even, there are $2^{m/2}$ palindromes among them. Indeed: the digits c_0 and c_m must be *ones*, but the digits $c_1, \ldots, c_{m/2}$ can be arbitrary; this yields the outcome $2^{m/2}$, because the remaining digits are now determined uniquely.

Similarly, if m is odd, then there are $2^{(m-1)/2}$ palindromes in that interval, owing to the freedom in the choice of the digits $c_1, \ldots, c_{(m-1)/2}$. Thus the number of palindromes smaller than 2^{11} equals

$$\sum_{m=0,2,4,6,8,10} 2^{m/2} + \sum_{m=1,3,5,7,9} 2^{(m-1)/2} = 94.$$

The (decimal) number 1988 has binary representation 11111000100. There are only two greater eleven-digit palindromes (with a *zero* or a *one* on the middle position, and with all *ones* on the remaining positions). Hence the conclusion: the condition $f(n) = n$ is satisfied by exactly 92 positive integers n, less than or equal to 1988.

1988/4

We replace the given numbers 70 and $5/4$ by an arbitrary positive integer n and an arbitrary positive real number c (the specific values of these parameters are of no help in solving the problem, while the presentation gains in simplicity in the general case). So we consider the function

$$f(x) = \frac{1}{x-1} + \frac{2}{x-2} + \cdots + \frac{n}{x-n} \tag{1}$$

defined on the set $J_0 \cup J_1 \cup \cdots \cup J_n$, where

$$J_0 = (-\infty, 1); \quad J_k = (k, k+1) \text{ for } k = 1, \ldots, n-1; \quad J_n = (n, \infty).$$

Within each one of these intervals the function f is continuous and strictly decreasing.

Fix $k \in \{1, \ldots, n-1\}$ and inspect the behavior of f at the endpoints of J_k: as x approaches k from the right, $f(x)$ goes to infinity; and when x approaches $k+1$ from the left, $f(x)$ goes to minus infinity (such is the behavior of the sum

$$\frac{k}{x-k} + \frac{k+1}{x-(k+1)};$$

and the remaining summands in the definition of f tend to finite limits). Thus there exists a unique point $\lambda_k \in J_k$ such that $f(\lambda_k) = c$.

The situation is not much different in the interval $J_n = (n, \infty)$. The limit of f at n is plus infinity, while the limit at $x \to \infty$ is 0. Since c is a positive number, we see that also in this case there is a unique point $\lambda_n \in J_n$ such that $f(\lambda_n) = c$.

We define the disjoint intervals $I_k = (k, \lambda_k]$ for $k = 1, \ldots, n$. Since f decreases in each J_k, it follows that

$$f(x) \geq c \quad \text{for } x \in I_k \quad \text{and} \quad f(x) < c \quad \text{for } x \in J_k \setminus I_k$$

(for $k = 1, \ldots, n$). In J_0 all the summands in (1) are negative, and hence $f(x) < 0$. Consequently,

$$\{x\colon\ f(x) \geq c\} = I_1 \cup I_2 \cup \cdots \cup I_n.$$

What remains to be done is to compute the sum of the lengths of the intervals $I_1, I_2, \ldots, I_n$. Consider the polynomial

$$Q(x) = (x-1)(x-2)(x-3)\cdots(x-n).$$

For $k \in \{1, \ldots, n\}$ let $P_k(x)$ be the polynomial obtained from $Q(x)$ by deleting the factor $(x-k)$. Define

$$\begin{aligned} F(x) &= Q(x)(c - f(x)) \\ &= cQ(x) - P_1(x) - 2P_2(x) - \cdots - nP_n(x). \end{aligned} \tag{2}$$

Each one of the numbers $\lambda_1, \ldots, \lambda_n$ is a root of the equation $f(x) = c$, hence also of $F(x) = 0$. There are no other roots, because F has degree n.

The coefficient of x^{n-1} in Q equals $-(1+2+\cdots+n)$. The coefficient of x^{n-1} in each of the polynomials $P_1, \ldots, P_n$ (of degree $n-1$) is equal to 1. In the polynomial F defined by (2) the term x^{n-1} appears therefore with the coefficient

$$D = -c(1+2+\cdots+n) - 1 - 2 - \cdots - n;$$

and the leading term x^n has coefficient c. The roots of F add up to

$$\lambda_1 + \lambda_2 + \cdots + \lambda_n = -\frac{D}{c} = \frac{c+1}{c} \cdot \frac{n(n+1)}{2}.$$

The interval I_k has length $\lambda_k - k$. Thus the sum of the lengths of $I_1, I_2, \ldots, I_n$ is equal to

$$\begin{aligned} (\lambda_1 - 1) + \cdots + (\lambda_n - n) &= (\lambda_1 + \cdots + \lambda_n) - (1 + \cdots + n) \\ &= \frac{c+1}{c} \cdot \frac{n(n+1)}{2} - \frac{n(n+1)}{2} = \frac{n(n+1)}{2c}. \end{aligned}$$

For $n = 70$, $c = 5/4$ this expression has value 1988, as claimed.

1988/5

First Solution. Assume $AB \leq AC$ without loss of generality. Let E and M be the mirror images of B and D (respectively) in the reflection across the bisector of angle BAD. Since $BD \perp AD$, it follows that $ME \perp AB$. Similarly, let F and N be the reflections of C and D across the bisector of angle CAD. Since $CD \perp AD$, we see that $NF \perp AC$. Clearly, $AM = AD = AN$.

Points A, D, E, F lie on the same straight line, in that order (in the extreme case of $AB = AC$, points E and F coincide). Ray ME cuts NF at a point P beyond E; note that $AMPN$ is a square.

The incircle of triangle ABD is also the incircle of triangle AEM; its center lies on the bisector of angle AME, i.e., on the diagonal MN of square $AMPN$. In the same way we show that the incenter of triangle ACD lies on the same diagonal. Therefore, according to the definition of K and L, point M coincides with K and point N coincides with L. Consequently

$$\begin{aligned} S &= \text{area}\,(ABC) = \text{area}\,(ABD) + \text{area}\,(ACD) \\ &= \text{area}\,(AEM) + \text{area}\,(AFN) \geq \text{area}\,(AMPN) \\ &= 2 \cdot \text{area}\,(AMN) = 2 \cdot \text{area}\,(AKL) = 2 \cdot T, \end{aligned}$$

as claimed.

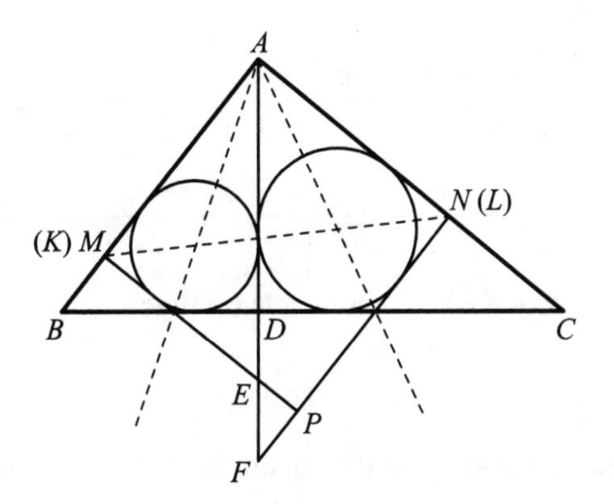

Second Solution. Assume that the triangle ABC is oriented counter-clockwise. The right triangles DAB and DCA are similar, in direct similitude. The incenter I of the first triangle corresponds in this similitude to the incenter J of the second triangle. Therefore triangle DIB is (directly) similar to triangle DJA.

Rays DI and DJ bisect the right angles BDA and ADC; hence $\angle IDA = \angle ADJ = 45°$. Consider the mapping f defined as counter-clockwise rotation by 45° about D, composed with dilatation, center D, in ratio DA/DJ. Thus f is a direct similitude with fixed-point D; it sends

J to A and I to B. It follows that line IJ is inclined to its f-image, i.e., line AB, under an angle of 45°. Hence AKL is an isosceles right triangle, $AK = AL$.

Triangles IAK and IAD are congruent (angles IAK and IAD are equal because AI is a bisector in triangle ABD, and angles AKI, ADI have size 45° each). So the segment AK (and hence also AL) has the same length as the altitude AD. Denote this length by h; it is well known that $h^2 = p \cdot q$, where $p = BD$, $q = CD$. Thus, finally,

$$\text{area}\,(ABC) = h \cdot \frac{p+q}{2} \geq h \cdot \sqrt{pq} = h^2 = AK \cdot AL = 2 \cdot \text{area}\,(AKL);$$

in other words, $S \geq 2 \cdot T$.

Remark. There are many methods of proving the result by calculation, involving the lengths of various segments, or trigonometry, or coordinate geometry, or a combination of these. The crucial fact (in all approaches) is that lines KL and AB make an angle of 45°. Readers fond of calculations can introduce the coordinate system so that $A = (0,0)$, $B = (b,0)$, $C = (0,c)$, verify that

$$I = \left(\frac{b\sqrt{b^2+c^2} - b^2 + bc}{2\sqrt{b^2+c^2}}, \frac{b^2c}{b^2+c^2+(b+c)\sqrt{b^2+c^2}}\right),$$

$$J = \left(\frac{bc^2}{b^2+c^2+(b+c)\sqrt{b^2+c^2}}, \frac{c\sqrt{b^2+c^2} - c^2 + bc}{2\sqrt{b^2+c^2}}\right),$$

and derive the assertion within reasonable (?) time.

1988/6

First Solution. According to the condition of the problem, the number $q = (a^2+b^2)/(ab+1)$ is a positive integer. It has to be proved that q is a perfect square.

The pair $(x, y) = (a, b)$ is one of the solutions of the equation

$$x^2 + y^2 = q(xy+1) \tag{1}$$

in nonnegative integers; thus the set of all such solutions is nonempty. Let (k, ℓ) be the pair of nonnegative integers satisfying equation (1) for which the sum $k + \ell$ is a minimum. Assume $k \leq \ell$; then $\ell > 0$ (since q is positive, the pair $(0,0)$ is *not* a solution of (1)). Rewrite equation (1) as

$$y(qx - y) = x^2 - q. \tag{2}$$

Consider the integer $j = qk - \ell$. Equation (2) is fulfilled by $x = k$, $y = \ell$, and therefore $\ell j = k^2 - q < k^2 \le \ell^2$, implying $j < \ell$.

Clearly, $\ell = qk - j$; the equality $j\ell = k^2 - q$ shows that equation (2) is also satisfied by $x = k$, $y = j$. And since $j < \ell$, the sum $k + j$ is smaller than $k + \ell$. This last number is (by definition) the least value of the sum $x + y$ for pairs of nonnegative integers satisfying (2).

It follows that j must be negative; so we have the inequality $qk < \ell$. Multiplying it by $1 \le -j$ we get $qk < -j\ell = q - k^2$. Thus $q(k-1) < -k^2$, and consequently $k - 1 < 0$. The only nonnegative integer for which this holds is $k = 0$. Setting in (1) $x = k = 0$, $y = \ell$, we eventually obtain $q = \ell^2$. So q is a perfect square, as required.

Second Solution. This is a geometric look at the problem. Let q be a fixed integer greater than 1 (see the Remark). We are going to examine the integer solutions of the equation

$$x^2 + y^2 = (xy + 1)q \tag{3}$$

and to prove that:

(a) if q is *not* a perfect square, equation (3) has no solution $(x, y) \in \mathrm{Z}^2$;

(b) if q *is* a perfect square, $q = \ell^2$, $\ell \in \mathrm{Z}^+$, then the recurrence formula

$$x_0 = 0, \quad x_1 = \ell, \quad x_{n-1} + x_{n+1} = qx_n \quad (n \in \mathrm{Z}), \tag{4}$$

defines a two-sided, strictly increasing sequence of integers, whose consecutive terms, in pairs ($x = x_n$, $y = x_{n+1}$ or vice versa; $n \in \mathrm{Z}$) constitute the complete solution of (3) in integers.

(The symbols Z, Z^+, Z^- and Z^2 denote, respectively, the set of all integers, the set of all positive integers, the set of all negative integers, and the set of all pairs (x, y) of integers—this last set being called the *integer lattice* on the coordinate plane.)

Evidently, claim (a) alone settles the assertion of the problem; however, it proves helpful to handle both claims, (a) and (b), simultaneously, and to examine the equation (3) on the integer lattice in its full generality. This approach sheds more light on the problem.

Rewriting equation (3) as

$$(q + 2)(x - y)^2 - (q - 2)(x + y)^2 = 4q$$

we see that it describes a hyperbola, with symmetry axes $y = x$, $y = -x$; when $q = 2$, the hyperbola degenerates to two parallel lines. Let H be the "upper" ($x < y$) branch of the hyperbola. Consider the point $Q_0 = (0, \sqrt{q})$ (remember that, *à priori*, the number $\sqrt{q}$ need not be an integer). Since Q_0 lies on H, the asymptotes of H (half-lines) form an obtuse angle and so H is the graph of a continuous, strictly increasing function f, mapping the set of all reals onto itself.

Choose a point $P_0 = (x_0, f(x_0)) \in H$ and consider the two-tailed sequence of iterates $x_n = f^n(x_0)$. (For $n \in \mathrm{Z}^+$, the symbol f^n denotes the n-fold composition of f; for $n = 0$, f^0 is the identity map; and for $n \in \mathrm{Z}^-$, the symbol f^n denotes the $|n|$-fold composition of the inverse function f^{-1}.) We will call the corresponding sequence of points

$$P_n = (x_n, x_{n+1}) = (x_n, f(x_n)) \in H$$

the *trajectory* passing through P_0. In particular, taking for P_0 the point $Q_0 = (0, \sqrt{q})$ we obtain the *main* trajectory; the predecessor of Q_0 in the trajectory order is $Q_{-1} = (-\sqrt{q}, 0)$.

Let now $\{P_n = (x_n, x_{n+1})\colon n \in \mathrm{Z}\}$ be any trajectory. Setting for x, y in (3) the coordinates of P_{n-1}, P_n and subtracting the resultant equalities we get the recurrence equation of (4), possibly with other initial conditions. (The initial data $x_0 = 0$, $x_1 = \sqrt{q}$ characterize the main trajectory.) The equation in (4) shows that if any one of the points $P_{n-1} = (x_{n-1}, x_n)$ and $P_n = (x_n, x_{n+1})$ is a lattice point, so is the other. Thus (use induction, forward and backward), every trajectory is either contained in Z^2 or disjoint from Z^2.

Note that equation (3) has no solutions in integers $x < 0 < y$. Hence, if H contains any lattice point P, then the trajectory through P, consisting entirely of lattice points, has no representative between Q_{-1} and Q_0. On the other hand, a trajectory cannot just skip the closed arc $Q_{-1}Q_0$ (this would require $f(x) > 0$ for some $x < -\sqrt{q}$, which is absurd). Thus the main trajectory is the only "candidate" for the set of lattice points on H.

In other words, equation (3) has solutions in Z^2 if and only if the main trajectory is composed of lattice points. This happens when $Q_0 \in \mathrm{Z}^2$, i.e., when q is a perfect square, in which case formula (4) with $\ell = \sqrt{q}$ defines the complete set of abscissae of lattice points on H; claims (a) and (b) are proved. (This sequence is symmetric with respect to 0; the other branch of the hyperbola is symmetric to H with respect to the origin. The lattice points on it are the same as on H, just with the roles of x and y interchanged; these points are also taken into account—this is provided by the words "vice versa" in the formulation of claim (6).)

Remark. The terms of the sequence (4) can be expressed by the explicit formula

$$x_n = \sqrt{\frac{q}{q^2-4}}\left(\left(\frac{q+\sqrt{q^2-4}}{2}\right)^n - \left(\frac{q-\sqrt{q^2-4}}{2}\right)^n\right), \qquad n \in \mathrm{Z}.$$

At the beginning we restricted attention to integers q greater than 1. For the case of $q = 1$, note that equation (3) is then equivalent to the following: $(x-y)^2+x^2+y^2 = 2$, with the obvious integer solutions $(0, \epsilon)$, $(\epsilon, 0)$ and (ϵ, ϵ), where $\epsilon = \pm 1$.

Thirtieth International Olympiad, 1989

1989/1

There are many ways to express this set in the way required. We will show one of them.

Let m be any positive odd integer and let k be any integer greater than 1. We are going to construct a rectangular array having m rows and k columns such that each column is a permutation of the set $\{1, 2, \ldots, m\}$ and all the row sums are equal.

If $k = 2$, we write in the first column all the numbers from 1 to m, arranged increasingly; in the second column we write the same numbers arranged decreasingly. The sum in each row will be equal to $m + 1$.

If $k = 3$, we compose an array putting in the ith row the numbers

$$\begin{array}{llll} [2i, & \frac{1}{2}(m+1)-i, & (m+1)-i] & \text{for } i < \frac{1}{2}m, \\ [2i-m, & \frac{1}{2}(3m+1)-i, & (m+1)-i] & \text{for } i > \frac{1}{2}m. \end{array}$$

Each one of the numbers $1, 2, \ldots, m$ appears exactly once on first position, exactly once on second and exactly once on third position. In each row the three numbers sum up to $\frac{3}{2}(m+1)$.

Finally, if $k > 3$, we write k as a sum of some 2s and 3s, and combine the two-column and three-column arrays constructed above.

So we have an array (matrix) of size $m \times k$, with the properties as promised. Now we add m to all numbers in the second column; and we add $2m$ to all numbers in the third column; and so on: to all numbers in the jth column we add $(j-1)m$. In the matrix thus modified, every number from 1 to $k \cdot m$ will show up exactly once.

Performing this construction for $m = 117$, $k = 17$, and defining A_i to be the set of elements in the ith row of this new matrix, we get all the conditions of the problem satisfied.

1989/2

(a) Let I be the incenter of triangle ABC. The bisector AA_1 of angle A bisects arc BC of the circumcircle. Thus BA_1C is an isosceles triangle, and we get

$$\angle CBA_1 = \angle BCA_1 = \frac{180^\circ - \angle BA_1C}{2} = \frac{\angle CAB}{2} = \angle IAB.$$

Therefore $\angle IBA_1 = \angle IBC + \angle CBA_1 = \angle ABI + \angle IAB = \angle A_1IB$, implying $A_1B = A_1I$.

In triangle ABC the external angle bisector at vertex B is perpendicular to the internal bisector. Thus IBA_0 is a right triangle. Point A_1 lies on its hypotenuse IA_0 and is equally distant from I and B; this means that it is the midpoint of IA_0. So the line BA_1 bisects the area of triangle IBA_0; and similarly, line CA_1 bisects the area of triangle ICA_0.

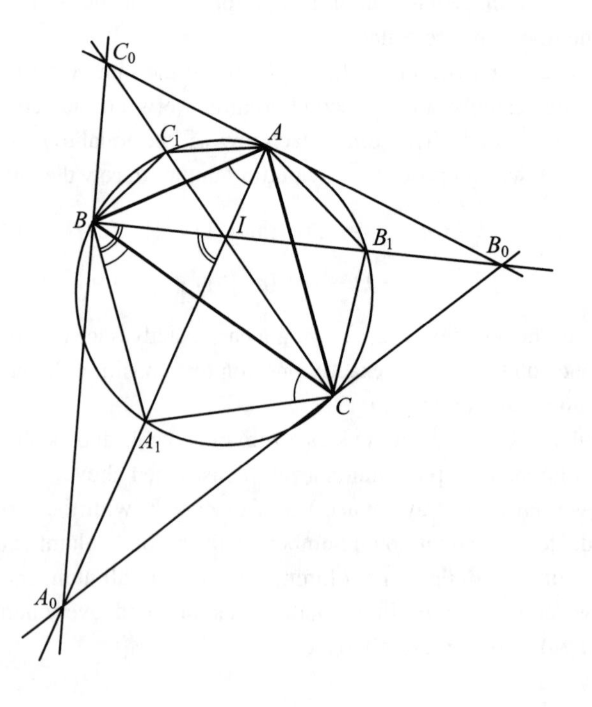

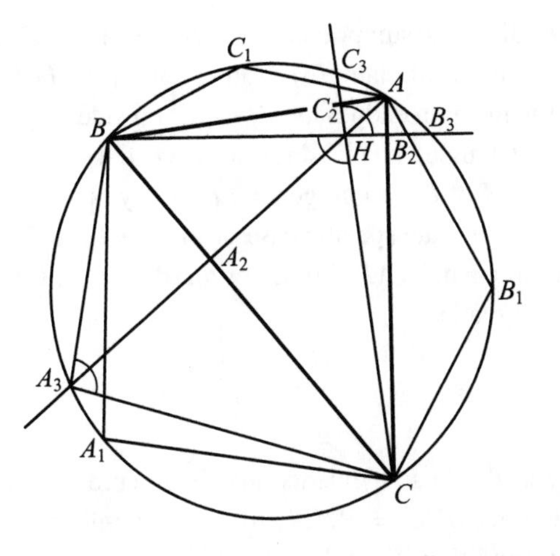

Consequently, the portion of triangle $A_0B_0C_0$ within sector BIC has area twice as large as the portion of hexagon $AC_1BA_1CB_1$ (in that sector). The analogous equality holds for the parts of those figures in the sectors CIA and AIB. The conclusion follows: the area of the whole triangle $A_0B_0C_0$ equals twice the area of the whole hexagon.

(b) Let H be the orthocenter of triangle ABC (the point at which the altitudes AA_2, BB_2, CC_2 concur). Extend those altitudes to their intersections with the circumcircle at A_3, B_3, C_3, respectively. The equality $\angle CHB = \angle B_2HC_2 = 180° - \angle CAB = \angle BA_3C$ shows that the points H and A_3 are symmetric across line BC. So the triangles BHC and BA_3C have equal areas. The area of triangle BA_3C does not exceed that of BA_1C because A_1 is the midpoint of arc BC. The inequality results:

$$\text{area}\ (HBA_1C) \geq \text{area}\ (HBC) + \text{area}\ (BA_3C) = 2 \cdot \text{area}\ (HBC).$$

Likewise,

$$\text{area}\ (HCB_1A) \geq 2 \cdot \text{area}\ (HCA), \quad \text{area}\ (HAC_1B) \geq 2 \cdot \text{area}\ (HAB).$$

Thus, according to the conclusion of part (a),

$$\begin{aligned} &\text{area}\ (A_0B_0C_0) \\ &\quad = 2 \cdot \text{area}\ (AC_1BA_1CB_1) \\ &\quad \geq 2 \cdot (2 \cdot \text{area}\ (HBC) + 2 \cdot \text{area}\ (HCA) + 2 \cdot \text{area}\ (HAB)) \\ &\quad = 4 \cdot \text{area}\ (ABC), \end{aligned}$$

proving the assertion of part (b)

Remark. Not all the assumptions about points A, B, C, A_0, B_0, C_0 are necessary for the proof of claim (b). Namely, if A_0A', B_0B', C_0C' are any concurrent segments with endpoints A', B', C' on sides B_0C_0, C_0A_0, A_0B_0, then the area of triangle $A'B'C'$ does not exceed 1/4 the area of triangle $A_0B_0C_0$. The proof of this more general property is somewhat more difficult, as compared with the specific case of lines AA_0, BB_0, CC_0 being the angle bisectors in triangle ABC (but not *too* difficult); we propose this as an exercise to the reader.

1989/3

Consider the set T whose elements are all ordered triples (M, N, P) of points of S such that $PM = PN$. Let t be the number of all such triples. From the given conditions (i) and (ii) we derive an upper bound and a lower bound for t.

For any two points $M, N \in S$ there exist in T at most two distinct triples with M and N being the first and the second entry. This follows from the fact that every point P completing the pair (M, N) to a triple (M, N, P) must lie on the perpendicular bisector of segment MN; according to condition (i), this bisector line passes through no more than two points of S. As there are $n(n-1)$ ordered pairs (M, N), we conclude that $t \leq 2n(n-1)$.

For any point $P \in S$ we can find at least $k(k-1)$ distinct triples in T whose last entry is P. This is a consequence of condition (ii): there are no fewer than k points of S equidistant from P; from these k points we can form $k(k-1)$ ordered pairs (M, N). Since we have n points P at our disposal, we get the lower estimate $t \geq nk(k-1)$.

The two estimates of t result in the inequality $2(n-1) \geq k(k-1)$; after recasting:

$$2n \geq k^2 - k + 2. \qquad (*)$$

The right-hand expression of this inequality strictly exceeds the number $(k - \frac{1}{2})^2$; hence the required result.

Remark. The inequality $(*)$ which we have worked out (just a trifle better than the claimed one) gives a lower estimate for the number of points in a set S satisfying the proposed conditions for a given integer $k \geq 1$. However, it is not clear for which integers k such a set S (of any cardinality) at all exists; if it does, its minimum cardinality may be very distant from what is provided by $(*)$.

It is not a trivial thing to devise any example of such a set S for a $k > 2$. Therefore we give an example for $k = 4$: the 8-element set S consisting of the vertices of a square and the vertices of equilateral triangles erected outwardly on the sides of the square; it does satisfy conditions (i) and (ii). (Note that inequality (∗) with $k = 4$ ensures only that $n \geq 7$, while the actual value of n in this example is 8.)

1989/4

Denote the radii: $r_1 = AD$, $r_2 = BC$. Let k_1 be the circle with center A and radius r_1, let k_2 be the circle with center B and radius r_2, and k_3 be the circle with center P and radius h. By the conditions of the problem, these three circles are pairwise externally tangent. Draw the line tangent to k_1 and k_2 at the respective points E and F, situated on the same side of line AB as P.

Line CD intersects circle k_1 at two points which we denote by D', D''; so D is one of them (in the limit case when CD touches k_1, these two points coincide). Fix labeling so that $PD' \leq PD''$. Analogously, define C' and C'' as the common points of line CD and circle k_2 (with $PC' \leq PC''$; thus $C = C'$ or $C = C''$). In each case, the circle k_3 lies in the quadrilateral $ABC'D'$, which in turn lies within $ABFE$.

Therefore k_3 is contained in the quadrilateral $ABFE$; and speaking more precisely—in the curvilinear triangle limited by pieces of k_1, k_2 and the segment EF.

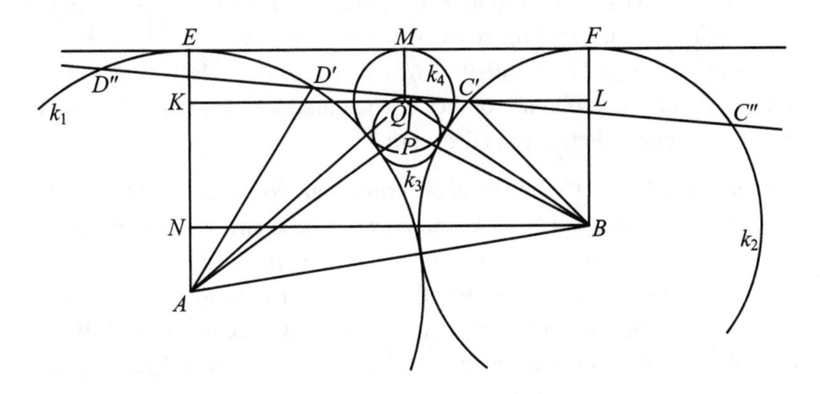

Consider the circle k_4 inscribed in this curvilinear triangle; assume it has center Q and radius r_4 (when line CD coincides with EF, circle k_3 coincides with k_4). Obviously, $h \leq r_4$. Denote by K, L, M the feet of perpendiculars from Q to lines AE, BF, EF respectively; and by N the foot of the perpendicular from B to line AE. From the right triangles AKQ, BLQ and BNA we compute

$$KQ = \sqrt{(r_1 + r_4)^2 - (r_1 - r_4)^2} = 2\sqrt{r_1 r_4},$$
$$QL = \sqrt{(r_2 + r_4)^2 - (r_2 - r_4)^2} = 2\sqrt{r_2 r_4},$$
$$BN = \sqrt{(r_1 + r_2)^2 - (r_1 - r_2)^2} = 2\sqrt{r_1 r_2}.$$

Since $KQ + QL = BN$, we obtain the equation $\sqrt{r_1 r_4} + \sqrt{r_2 r_4} = \sqrt{r_1 r_2}$. Division by $\sqrt{r_1 r_2 r_4}$ now implies:

$$\frac{1}{\sqrt{r_1}} + \frac{1}{\sqrt{r_2}} = \frac{1}{\sqrt{r_4}} \le \frac{1}{\sqrt{h}};$$

and this is the inequality we were about to prove.

1989/5

First Solution. Fix n and define $N = 1 + ((n+1)!)^2$. We will show that the numbers $N+1, N+2, \ldots, N+n$ have the demanded property: no one of them is an integral power of a prime number.

Assume to the contrary that for some integers k, m $(1 \le m \le n)$ and for a certain prime number p the equality $N+m = p^k$ is fulfilled; obviously $k \ge 0$ (otherwise p^k would not be an integer).

The number $m+1$ is a divisor of $(n+1)!$, hence also of the number $((n+1)!)^2 + (m+1) = N+m = p^k$. Since $1 < m+1 < N+m$, we get that $m+1$ is a number of the form p^r with $0 < r < k$.

Thus p^{r+1} is a divisor of $p^k = N+m$ and of $((n+1)!)^2 = N-1$; so it has to divide the difference of these two numbers, equal to p^r. But this is impossible; contradiction ends the proof.

Second Solution. The result is an immediate consequence of the Chinese remainder theorem (see *Glossary*): we choose any distinct primes $p_1, p_2, \ldots, p_{2n}$ and define $a_i = p_{2i-1}p_{2i}$ and $r_i = a_i - i$ for $i = 1, \ldots, n$. According to the theorem just mentioned, there exists an integer N such that $N \equiv r_i \pmod{a_i}$ for $i = 1, \ldots, n$. We may assume $N > 0$ (if not, we just add to N a sufficiently large multiple of $a_1 a_2 \cdots a_n$). Then there are integers $q_1, \ldots, q_n$ such that

$$N + i = (a_i q_i + r_i) + i = a_i(q_i + 1) = p_{2i-1}p_{2i}(q_i + 1) \quad \text{for } i = 1, \ldots, n.$$

Thus each one of the numbers $N+1, N+2, \ldots, N+n$ has at least two distinct prime divisors, and hence cannot be an integral power of any prime.

1989/6

First Solution. A permutation with property P will be called *nice* for brevity (in fact, this was the term used in discussions at the 1989 IMO Jury, before it was replaced, in the official formulation, by the ponderous *property P*).

The claim is obvious for $n = 1$. Assume $n \geq 2$ for the sequel. To stick to this terminological style, let us call a permutation $(x_1, x_2, \ldots, x_{2n})$ of the set $\{1, 2, \ldots, 2n\}$ *very nice* if the equality $|x_i - x_{i+1}| = n$ holds for exactly one $i \in \{2, \ldots, 2n-1\}$ and does *not* hold for $i = 1$.

Let now $(x_1, x_2, \ldots, x_{2n})$ be a very nice permutation and let ℓ be the only index greater than 1 for which the equality $|x_\ell - x_{\ell+1}| = n$ is satisfied. Move the element x_ℓ to the initial position and relabel the terms; i.e., define

$$y_i = \begin{cases} x_\ell & \text{for} \quad i = 1, \\ x_{i-1} & \text{for} \quad i = 2, \ldots, \ell, \\ x_i & \text{for} \quad i = \ell + 1, \ldots, 2n. \end{cases}$$

The resulting permutation $(y_1, y_2, \ldots, y_{2n})$ is not nice: no two neighboring terms differ exactly by n.

And conversely, if $(z_1, z_2, \ldots, z_{2n})$ is a non-nice permutation, we can move the element z_1 to the position immediately preceding the only term z_m satisfying the equality $|z_1 - z_m| = n$ (note that there is exactly one such number in the set $\{1, 2, \ldots, 2n\}$). The resulting permutation is very nice.

These two operations, mutually inverse, establish a one-to-one correspondence between the set of all very nice permutations and the set of all non-nice permutations. So these two sets are equipotent. And since there exist nice permutations which are not very nice, it follows that the nice permutations are more numerous than those that are not nice.

Second Solution. To every number $x \in \{1, 2, \ldots, 2n\}$ there corresponds a unique $y \in \{1, 2, \ldots, 2n\}$ such that $|x - y| = n$; such numbers x, y will be called *twins*. A permutation of the set $\{1, 2, \ldots, 2n\}$ is nice (has property P) if at least one pair of twins appear as neighbors; it is non-nice if there is no such pair. A permutation will be called *singly-nice* if exactly one pair of twins appear as neighbors. We are going to show that the singly-nice permutations alone suffice to outweigh the non-nice ones; this of course will be enough for the solution of the problem.

Denote by $f_0(n)$ the number of all non-nice permutations of the set $\{1, 2, \ldots, 2n\}$ and by $f_1(n)$ the number of all singly-nice permutations. Our intended task is to show that $f_1(n) > f_0(n)$.

We will derive two recursion formulas. Consider a non-nice permutation $(x_1, x_2, \ldots, x_{2n})$. Remove the element x_{2n} together with its twin. What remains is a permutation of $2n-2$ elements, either non-nice or singly-nice; the latter case occurs if the removed element (the twin of x_{2n}) was separating a pair of twins. Clearly, x_{2n} can take $2n$ values and, in the first case, its twin can take any one of $2n-2$ positions, whereas in the second case its position is determined by the resulting singly-nice permutation of $2n-2$ elements. The first recursion formula follows:

$$f_0(n) = 2n((2n-2)f_0(n-1) + f_1(n-1)).$$

Now, let $(x_1, x_2, \ldots, x_{2n})$ be a singly-nice permutation and let (x_j, x_{j+1}) be the unique neighboring twin pair. Remove this pair to obtain a permutation of $2n-2$ elements. Again, the new permutation will be either non-nice or singly-nice; the latter case occurs when the removed pair was previously separating another pair of twins. The pair (x_j, x_{j+1}) is chosen out of n twin pairs and can be arranged in two ways. In the first case, this pair can take any one of $2n-1$ positions, whereas in the second case its position is determined by the resulting singly-nice permutation of $2n-2$ elements. The second recursion formula follows:

$$f_1(n) = 2n((2n-1)f_0(n-1) + f_1(n-1)).$$

Subtracting the first recursion formula from the second we get

$$f_1(n) - f_0(n) = 2nf_0(n-1).$$

The promised inequality $f_1(n) > f_0(n)$ results for $n \geq 3$. (The case of $n = 2$ is an exception, as $f_0(1) = 0$ and so $f_1(2) = f_0(2)$; here we must use the fact that there exist other nice permutations of $\{1, 2, 3, 4\}$, in addition to those which are singly-nice.)

Third Solution. For $k = 1, 2, \ldots, n$, let $\mathcal{P}_k$ be the set of those permutations of the set $\{1, 2, \ldots, 2n\}$ in which the "twin" numbers k and $k+n$ appear on neighboring positions. The union $\mathcal{P} = \mathcal{P}_1 \cup \cdots \cup \mathcal{P}_n$ is precisely the set of all nice permutations. The cardinality of this union is expressed by the inclusion-exclusion formula

$$|\mathcal{P}| = c_1 - c_2 + c_3 - c_4 + \cdots + (-1)^{n-1}c_n$$

where

$$c_r = \sum_{1 \le k_1 < \cdots < k_r \le n} |\mathcal{P}_{k_1} \cap \cdots \cap \mathcal{P}_{k_r}|.$$

To evaluate these summands, fix an $r \in \{1, 2, \ldots, 2n\}$ and choose r numbers $k_1 < \cdots < k_r$ from $\{1, 2, \ldots, n\}$. How many permutations are there that belong simultaneously to the sets $\mathcal{P}_{k_1}, \ldots, \mathcal{P}_{k_r}$?

The selected numbers $k_1, \ldots, k_r$ determine r twin pairs, the elements in each pair having to occupy neighboring positions. Regarding each one of those pairs as a "brick," we have $(2n - r)!$ ways to arrange the $2n - r$ objects: the r "bricks" and the remaining $2n - 2r$ elements of $\{1, 2, \ldots, 2n\}$. This number has to be multiplied by 2^r, as there are two ways, within each "brick," to arrange a twin pair $\{k, k + n\}$ into an ordered pair.

Since we have

$$\binom{n}{r}$$

possibilities of choosing the numbers $k_1, \ldots, k_r$, we arrive at the value

$$c_r = \binom{n}{r} \cdot (2n - r)! \cdot 2^r.$$

A simple calculation shows that, for $r = 1, \ldots, n - 1$,

$$\frac{c_{r+1}}{c_r} = \frac{2n - 2r}{2n - r} \cdot \frac{1}{r + 1} < 1 \cdot \frac{1}{2} = \frac{1}{2};$$

hence $c_{r+1} < c_r/2 < c_r$. For $r = 1$ we have $c_1 = (2n)!$. Thus, finally,

$$|\mathcal{P}| = (c_1 - c_2) + (c_3 - c_4) + \cdots + \begin{cases} (c_{n-1} - c_n), & n \text{ even}, \\ c_n, & n \text{ odd}, \end{cases}$$

$$\ge c_1 - c_2 > c_1 - \frac{1}{2} \cdot c_1 = \frac{1}{2} \cdot c_1 = \frac{1}{2} \cdot (2n)!$$

showing that the nice permutations make up more than a half of all the $(2n)!$ permutations of $\{1, 2, \ldots, 2n\}$, and consequently are more numerous than those that are not nice.

Remark. It can be shown that the number of non-nice permutations, divided by $(2n)!$, tends to the limit $1/e$; readers can consider this as an exercise and derive the limit behavior from the equations in the third solution.

Thirty-first International Olympiad, 1990

1990/1

Draw the segments from D to A, B and M. Denote by Ω the circle passing through the points A, B, C, D, and by ω the circle passing through D, E and M. This latter circle enters the interior of the convex angle BEC; thus F (the point where the tangent to ω at E cuts line BC) lies between B and C. In particular, F and D are on distinct sides of line AB, and so $\angle MEF = \angle MDE$, by the tangent-chord theorem. Therefore

$$\angle BEF = \angle MEF = \angle MDE < \angle BDE = \angle BDC = \angle BAC.$$

This inequality implies that G (the point where the tangent FE cuts line AC) lies on ray CA beyond A. Thus, again by the tangent-chord theorem,

$$\angle CEF = \angle DEG = \angle EMD = \angle AMD, \tag{1}$$

whence

$$\angle CEG = 180° - \angle CEF = 180° - \angle AMD = \angle BMD. \tag{2}$$

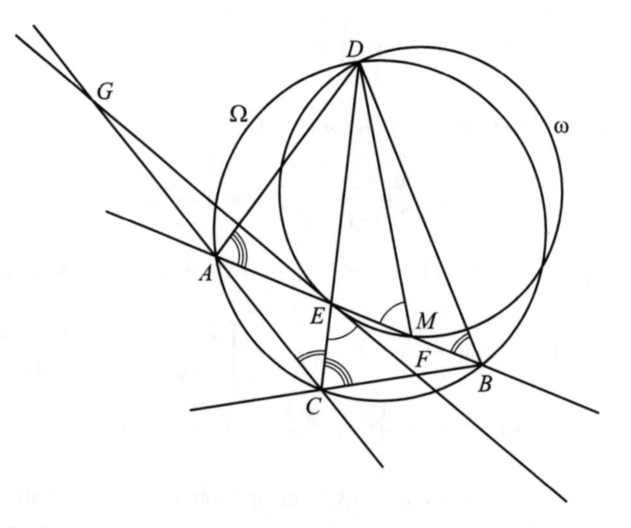

Since the points A, C, B, D lie on Ω in that order, the following angles are equal:

$$\angle MAD = \angle BAD = \angle BCD = \angle ECF \tag{3}$$

and

$$\angle MBD = \angle ABD = \angle ACD = \angle ECG. \tag{4}$$

In view of (1) and (3), triangles MAD and ECF are similar; and by (2) and (4), triangles MBD and ECG are similar. Consequently,

$$\frac{EG}{EF} = \frac{EG}{MD} \cdot \frac{MD}{EF} = \frac{EC}{MB} \cdot \frac{MA}{EC} = \frac{MA}{MB}$$
$$= \frac{t \cdot AB}{(1-t) \cdot AB} = \frac{t}{1-t};$$

and this is the ratio we had to evaluate.

1990/2

Label the points of E, in cyclic order, $P_0, P_1, \ldots, P_{2n-2}$; the position of P_0 and the orientation of the circle are arbitrary. Take any points $P_i, P_j \in E$. One of the arcs with endpoints P_i, P_j contains $|i-j|-1$ points from E and the other one contains $(2n-2) - |i-j|$ points. We wish to inspect those pairs (P_i, P_j) for which one of these two numbers equals n; i.e., such that one of the two alternatives holds: either $|i-j| = n+1$ or $|i-j| = n-2$. For $i, j \in \{0, 1, \ldots, 2n-2\}$ this is equivalent to:

$$|i-j| \equiv \pm(n-2) \pmod{2n-1}. \tag{1}$$

Let us color red all segments P_iP_j for which condition (1) is satisfied. Thus every point of E is connected by red segments with exactly two other points at distance $n-2$ from it (measured in "unit" arcs). The problem reduces to the following: find the smallest integer k such that every k-element subset of E contains at least one pair of points linked by a red segment.

Consider the sequence $i_0, i_1, i_2, \ldots$ of indices $i_r \in \{0, 1, \ldots, 2n-2\}$ defined by the recursion

$$i_r \equiv i_{r-1} + (n-2) \pmod{2n-1}, \tag{2}$$

with i_0 chosen arbitrarily; this recursive construction is continued until a repetition appears. The segments $P_{i_{r-1}}P_{i_r}$ form a self-intersecting (star-shaped) closed polygonal line. The sequence (2) will contain every integer from 0 to $2n-2$ if and only if the numbers $2n-1$ and $n-2$ are relatively prime; and this is the case if and only if $2n-1$ is not divisible by 3. Then the resulting polygon L has $2n-1$ vertices and the same number of edges; all the edges are red, and there are no other red segments.

In the remaining case, when $2n - 1$ is divisible by 3, the sequence (2) will only contain those indices that leave the same remainder in division by 3 as i_0 does. The broken line $P_{i_0} P_{i_1} P_{i_2} \ldots$ will close after $(2n - 1)/3$ steps. Denote it by L_0. Starting from P_{i_0+1} or P_{i_0+2} (rather than from P_{i_0}) we can analogously form two other closed polygons, L_1 and L_2. The sets of vertices of L_0, L_1, L_2 are disjoint and their union is the whole set E; all the edges are red, and there are no other red segments.

In the first case (when $2n - 1$ is not divisible by 3), the minimum value of k such that every k-element set of "black" points contains the endpoints of at least one red segment is $k = n$. Indeed: no matter how we choose n "black" points, there will be two adjacent vertices of the $(2n - 1)$-lateral polygon L among them; and if k is smaller than n, we are able to blacken k vertices of L, no two of them being adjacent.

In the second case ($2n - 1$ divisible by 3), we have $2n - 1 = 6m + 3$ for a certain integer m. Each one of the three polygons L_0, L_1, L_2 has $2m + 1$ vertices; we can blacken m vertices of each, a total of $3m$, no two being adjacent. But we could not avoid an adjacent pair while trying to blacken $3m + 1$ points. Thus in this case the minimum k is $k = 3m + 1 = n - 1$.

Hence the outcomes: the smallest value of k for which every such coloring of k points is *good* equals:

$$\begin{cases} n & \text{when } 2n - 1 \text{ is not divisible by 3,} \\ n - 1 & \text{when } 2n - 1 \text{ is divisible by 3.} \end{cases}$$

1990/3

Suppose an integer $n > 1$ satisfies the given condition; obviously, n cannot be even. Write n in the form $n = 3^k m$ with integers $k \geq 0$, $m \geq 1$, where m is not divisible by 2 or 3.

For any integer $k > 0$ and any real number x we have

$$x^{3^k} + 1 = \left(x^{3^{k-1}}\right)^3 + 1 = \left(x^{3^{k-1}} + 1\right)\left(x^{2\cdot 3^{k-1}} - x^{3^{k-1}} + 1\right).$$

If also $k - 1 > 0$, we can analogously factor the sum $(x^{3^{k-1}} + 1)$. After k steps we arrive at the equality

$$x^{3^k} + 1 = (x + 1)\prod_{j=0}^{k-1}\left(x^{2\cdot 3^j} - x^{3^j} + 1\right),$$

valid also for $k = 0$ (in which case the "void" product that appears on the right side represents the number 1). Setting $x = 2^m$ we get

$$2^n + 1 = (2^m + 1) \prod_{j=0}^{k-1} q_j \quad \text{where} \quad q_j = 4^{3^j m} - 2^{3^j m} + 1. \tag{1}$$

Let us examine the remainders of numbers 2^ℓ in division by 9, for odd exponents ℓ. Since $2^1 \equiv 2$, $2^3 \equiv 8$, $2^5 \equiv 5$, $2^7 \equiv 2 \pmod 9$, the residues 2, 8, 5 will repeat cyclically, and we easily obtain

$$4^\ell - 2^\ell + 1 \equiv 3 \pmod 9 \quad \text{for } \ell = 1, 3, 5, 7, 9, \ldots. \tag{2}$$

Since m is not divisible by 3, the residue of 2^m can be 2 or 5, and not 8. So $2^m + 1 \equiv \pm 3 \pmod 9$. Setting in (2) $\ell = 3^j m$ we see that $q_j \equiv 3 \pmod 9$ for $j = 0, \ldots, k-1$. Therefore the product on the right side of (1) is divisible by 3^{k+1}, but not by 3^{k+2}.

According to the condition of the problem, the left side of (1) represents a number divisible by n^2, hence also by 3^{2k}. It follows that $2k \le k+1$, which means that k is either 0 or 1, and so

$$n = m \quad \text{or} \quad n = 3m. \tag{3}$$

We claim that $m = 1$. Assume, to the contrary, that $m > 1$ and let p be the smallest prime divisor of m. Since m is not divisible by 2 or 3, the prime p is at least 5. Obviously, $p - 1$ is coprime to m.

Using once more the condition of the problem, we have that $2^n + 1$ is divisible by n, hence also by p. Thus $2^n \equiv -1 \pmod p$ and by squaring $2^{2n} \equiv 1 \pmod p$. Also, by Fermat's little theorem (see *Glossary*), $2^{p-1} \equiv 1 \pmod p$.

Let d be the greatest common divisor of $2n$ and $p - 1$, and let α be the smallest positive integer such that $2^\alpha \equiv 1 \pmod p$. Dividing $p - 1$ by α we get $p - 1 = q\alpha + r$ with $0 \le r < \alpha$. Then $2^r \equiv 2^{p-1} \equiv 1 \pmod p$. By the definition of α, the exponent r must be 0. This means that $p - 1$ is divisible by α.

From $2^{2n} \equiv 1 \pmod p$ we analogously obtain that $2n$ is divisible by α. Therefore α is a divisor of d. Since $2^\alpha \equiv 1 \pmod p$, we conclude that $2^d \equiv 1 \pmod p$.

Note that d, a divisor of $2n$, is in view of (3) a divisor of $6m$. Also, d divides the number $p - 1$, which is coprime to m. Thus d is a divisor of 6, i.e., one of the numbers 1, 2, 3, 6.

Now it follows from $2^d \equiv 1 \pmod{p}$ that p is a prime divisor of one of the numbers 1, 3, 7, 63. Since $p \geq 5$, the only possibility is $p = 7$. Accordingly, the number $2^n + 1$ (divisible by n, by the condition of the problem) must be divisible by 7. This is, however, impossible, as the powers of 2 leave only remainders 1, 2 and 4 in division by 7. Contradiction proves our claim: $m = 1$. And since we are only considering integers $n > 1$, it follows from (3) that n has to be equal to 3.

Evidently, $n = 3$ satisfies the given condition; so it is the unique solution to the problem.

1990/4

Suppose f is a function satisfying the given functional equation. The usual start to such problems is to try some typical substitutions. Denote $f(1)$ by c. Setting $x = 1$ in the equation we get

$$f(f(y)) = \frac{c}{y}. \tag{1}$$

It follows that f is a bijective mapping of Q^+ onto (the whole of) Q^+. With $y = 1$, equation (1) gives $f(c) = c$. Then set in (1) $y = c$, to obtain $f(f(c)) = 1$; hence $c = 1$.

Take any $x, y \in \mathrm{Q}^+$. Since f is surjective, we can write $y = f(z)$ for a certain $z \in \mathrm{Q}^+$. Applying the given equation (with z in place of y), and using (1) with $c = 1$, we obtain

$$f(xy) = f(xf(z)) = \frac{f(x)}{z} = f(x) \cdot \frac{1}{z} = f(x) \cdot f(f(z)) = f(x)f(y).$$

Thus we have shown that the given functional equation implies the system of two equations

$$f(xy) = f(x)f(y), \qquad f(f(x)) = \frac{1}{x} \quad \text{for } x, y \in \mathrm{Q}^+. \tag{2}$$

And conversely, it is immediate that if a function $f\colon \mathrm{Q}^+ \to \mathrm{Q}^+$ fulfils this system, then it satisfies the original equation.

Let $p_1 = 2$, $p_2 = 3$, $p_3 = 5, \ldots$ be the increasing sequence of all primes. Every number $x \in \mathrm{Q}^+$ different from 1 can be uniquely written in the form

$$x = \prod_{i=1}^{n} p_i^{k_i}, \tag{3}$$

where n is a positive integer and the exponents $k_1, \ldots, k_n$ are integers (not necessarily positive), with $k_n \neq 0$. In view of equations (2) and representation (3), the function f is fully determined by its values on the primes.

Let $q_1, q_2, q_3 \ldots$ be any sequence of distinct positive rationals. Imposing the values $f(p_i) = q_i$ for $i = 1, 2, 3, \ldots$ (and $f(1) = 1$) we obtain the function

$$f(x) = \prod_{i=1}^{n} q_i^{k_i} \quad \text{for } x \text{ as in (3).} \tag{4}$$

It satisfies the first equation of (2). To ensure that the second equation be fulfilled, it suffices to require that the equality $f(f(p)) = 1/p$ should hold for every prime number p.

This can be achieved in many ways. For instance: let

$$q_i = \begin{cases} p_{i+1} & \text{for } i \text{ odd,} \\ 1/p_{i-1} & \text{for } i \text{ even,} \end{cases}$$

and let f be defined by (4). Then

$$f(f(p_i)) = f(q_i) = \begin{cases} f(p_{i+1}) = q_{i+1} = 1/p_i & \text{for } i \text{ odd,} \\ f(1/p_{i-1}) = 1/q_{i-1} = 1/p_i & \text{for } i \text{ even.} \end{cases}$$

So $f(f(p)) = 1/p$ for all primes p, and consequently the function f thus defined satisfies both equations of the system (2); hence it satisfies the equation from the problem statement.

1990/5

Let us first look for a beneficial strategy from the view-point of player A. If the initial n_0 satisfies the inequality $45 \leq n_0 \leq 1990$, player A chooses $n_1 = 1990$ and wins.

If $27 \leq n_0 \leq 44$, player A chooses $n_1 = 720 = 2^4 \cdot 3^2 \cdot 5$. The smallest number B can choose is $n_1/2^4 = 45$; and the greatest is $n_1/2 = 360$. So $45 \leq n_2 \leq 360$ and A wins by choosing $n_3 = 1990$.

If $15 \leq n_0 \leq 26$, A takes $n_1 = 210 = 2 \cdot 3 \cdot 5 \cdot 7$. The number n_2 chosen by B must satisfy $n_1/7 = 30 \leq n_2 \leq n_1/2 = 105$, giving a winning position to A ($n_3 = 720$, $n_5 = 1990$, or $n_3 = 1990$ at once).

If $11 \leq n_0 \leq 14$, A can choose $n_1 = 105 = 3 \cdot 5 \cdot 7$. Then B must choose n_2 so that $15 \leq n_2 \leq 35$, again providing winning position to A ($n_3 = 210$ or 720 etc.).

If $8 \leq n_0 \leq 10$, A chooses $n_1 = 60 = 2^2 \cdot 3 \cdot 5$, forcing $12 \leq n_2 \leq 30$; then A takes $n_3 = 105$ (or 210, or 720) and wins.

For n_0 smaller than 8 this method will not work. So let us look at numbers greater than 1990.

Assume that the number n_{2k} given by B (at any moment of the game) exceeds 1990. Player A can find in the interval $[n_{2k}, 2n_{2k})$ a number of the form $9 \cdot 2^j$ and choose it to be n_{2k+1}. The next number chosen by B will be not greater than $9 \cdot 2^{j-1}$ (hence smaller than n_{2k}), but not less than 9.

Thus if $n_0 > 1990$, player A forces B to choose smaller and smaller numbers, without descending below 9. Sooner or later, A will receive from B a number from the interval [9, 1990] and will be able to apply the strategy described in the previous paragraphs.

Now consider small initial values. For $n_0 = 2, 3, 4, 5$ player A must choose an n_1 not exceeding 25, hence having at most two distinct prime factors. Dividing it by one of those factors, in a possibly high power, player B can get a number $n_2 < \sqrt{n_1} \leq n_0$. If $n_2 > 1$, then B can in the same way have in the next step a number $n_4 < n_2$, and soon will reach 1 (and victory).

It remains to consider $n_0 = 6$ and $n_0 = 7$. The only choices that prevent A from an immediate loss of the game are $n_1 = 30$ and $n_1 = 42$ (these are the only numbers not greater than 49 that have three distinct prime factors). Now B chooses $n_2 = 6$ or $n_2 = 7$ and the cycle repeats. Nobody wins.

The answer follows: player A has a winning strategy for $n_0 \geq 8$; player B has a winning strategy for $n_0 \leq 5$. For $n_0 = 6, 7$ neither player has a winning strategy.

1990/6

The complex plane is an appropriate environment to handle this problem. We are looking for a sequence $(n_0, n_1, \ldots, n_{1989})$ that is a permutation of the set $\{0, 1, \ldots, 1989\}$ and satisfies the equation

$$\sum_{r=0}^{1989} (1 + n_r)^2 (\cos r\varphi + i \sin r\varphi) = 0 \quad \text{where } \varphi = \frac{2\pi}{1990}. \tag{1}$$

For suppose $(n_0, n_1, \ldots, n_{1989})$ is such a sequence. For each consecutive $r \in \{0, 1, \ldots, 1989\}$ let $\mathbf{v}_r$ be the vector represented by the complex number that occurs in (1) as the rth summand; moreover, let $\mathbf{v}_{1990} = \mathbf{v}_0$. Attach $\mathbf{v}_0$ at any point of the complex plane and construct from the vectors $\mathbf{v}_r$ a closed polygonal line, attaching each consecutive vector at the endpoint of the preceding one; the line closes because the sum of the $\mathbf{v}_r$s is the zero vector, according to equation (1).

The oriented angle between any two adjacent vectors $\mathbf{v}_{r-1}$, $\mathbf{v}_r$ has size φ; consequently, the resulting broken line describes a convex 1990-gon with all its angles equal. The lengths of its sides, i.e., the lengths of the vectors $\mathbf{v}_r$, are the numbers $(1+n_0)^2, \ldots, (1+n_{1989})^2$.

All that remains is to find the integers n_r. Let $\epsilon = \cos\varphi + i\sin\varphi$; then by de Moivre's formula, $\epsilon^r = \cos r\varphi + i\sin r\varphi$ for each integer r (in particular, $\epsilon^{1990} = 1$). The required equality (1) rewrites as

$$\sum_{r=0}^{1989}(1+n_r)^2\epsilon^r = 0. \tag{2}$$

The crucial fact for the further reasoning is that the number 1990 factors into the product of three primes: $1990 = 2\cdot 5\cdot 199$. Define:

$$\alpha = \epsilon^{2\cdot 5}, \qquad \beta = \epsilon^{2\cdot 199}, \qquad \gamma = \epsilon^{5\cdot 199};$$

then $\alpha^{199} = \beta^5 = \gamma^2 = 1$; $\alpha, \beta, \gamma \neq 1$, and hence

$$\sum_{j=0}^{198}\alpha^j = 0, \qquad \sum_{k=0}^{4}\beta^k = 0, \qquad \sum_{l=0}^{1}\gamma^l = 0.$$

(Clearly, $\gamma = -1$; but the presentation gains in clarity if we leave the "general" symbol γ.)

Let T be the set of all ordered triples (j,k,l) of integers $0 \le j < 199$, $0 \le k < 5$, $0 \le l < 2$. There are 1990 such triples. Note that if $g(j,k,l)$ is a function defined on T, independent of one of the variables, then

$$\sum_{(j,k,l)\in T} g(j,k,l)\alpha^j\beta^k\gamma^l = 0. \tag{3}$$

Indeed; assume that g does not depend, say, on k; i.e., $g(j,k,l) = h(j,l)$. Then the sum in (3) equals

$$\sum_{j=0}^{198}\alpha^j\sum_{l=0}^{1}\gamma^l h(j,l)\sum_{k=0}^{4}\beta^k;$$

and this is zero, as $\sum_{k=0}^{4}\beta^k = 0$.

Evidently, the numbers of the form

$$10j + 2k + l, \qquad (j,k,l)\in T, \tag{4}$$

are all different and sweep the whole set $\{0, 1, \ldots, 1989\}$. We now show that also the remainders in division by 1990 left by the numbers

$$2\cdot 5\cdot j + 2\cdot 199\cdot k + 5\cdot 199\cdot l, \qquad (j,k,l)\in T, \tag{5}$$

have the same property. Since T is a 1990-element set, it suffices to show that those remainders are pairwise different.

Thus suppose that the numbers (5) defined by two triples (j,k,l) and (j',k',l') from T are congruent modulo 1990; then the difference $j-j'$ is divisible by 199, the difference $k-k'$ is divisible by 5, and the difference $l-l'$ is divisible by 2; and this means, by the definition of T, that $j=j'$, $k=k'$, $l=l'$. So, indeed, the numbers (5) (reduced modulo 1990) range without repetition over the set $\{0,1,\ldots,1989\}$.

The following definition is now legitimate: for any integer r from the set $\{0,1,\ldots,1989\}$ let (j,k,l) be the unique triple from T for which the number in (5) gives remainder r in division by 1990; define n_r to be the value of the sum in (4) for this specific triple (j,k,l): $n_r = 10j+2k+l$. In view of the foregoing observations, the sequence $(n_0, n_1, \ldots, n_{1989})$ thus obtained is a permutation of the set $\{0,1,\ldots,1989\}$. If we show that equation (2) is satisfied with these numbers n_r, the result will be proved.

Take an r and let j, k and l be as specified. Since the expression in (5) is congruent to r modulo 1990, and since $\epsilon^{1990}=1$, we have

$$\epsilon^r = \epsilon^{2\cdot5\cdot j}\epsilon^{2\cdot199\cdot k}\epsilon^{5\cdot199\cdot l} = \alpha^j\beta^k\gamma^l.$$

As r runs from 0 to 1989, the triple (j,k,l) ranges over the whole set T. According to the definition of n_r, the left side expression in (2) has value

$$\sum_{(j,k,l)\in T} (1+10j+2k+l)^2\alpha^j\beta^k\gamma^l.$$

Expanding the square we obtain the sum of the squares of the four terms (in the parentheses) plus the doubled sum of "mixed products" (like $10j\cdot2k$); anyway, we get a sum of terms, each of which involves no more than two symbols out of j,k,l. As noticed earlier, equation (3), summation over all triples $(j,k,l)\in T$ produces value 0, as desired. Equation (2) is fulfilled and the solution is complete.

Thirty-second International Olympiad, 1991

1991/1

Line CI is the bisector of angle C in triangle ACA', and BI is the bisector of angle B in triangle ABA'. Therefore I divides segment AA' in the ratio $A'I/AI = A'C/AC = A'B/AB$. Denoting by a, b, c the lengths of the sides BC, CA, AB we hence obtain

$$a = A'C + A'B = (b+c)\cdot\frac{A'I}{AI}.$$

Consequently

$$\frac{AA'}{AI} = \frac{AI + A'I}{AI} = 1 + \frac{A'I}{AI} = 1 + \frac{a}{b+c} = \frac{a+b+c}{b+c},$$

and we have analogous expressions for BB'/BI and CC'/CI. The problem reduces to showing that

$$\frac{1}{4} < \frac{b+c}{a+b+c}\cdot\frac{c+a}{a+b+c}\cdot\frac{a+b}{a+b+c} \le \frac{8}{27} \tag{1}$$

for a, b, c sides of a triangle.

These are very standard cyclic inequalities; there are several (equally standard) methods of proof. For instance, readers familiar with convex functions and the technique of Jensen's inequality (see *Glossary*) may try that approach. We show a more elementary way. The substitution

$$x = \frac{-a+b+c}{a+b+c},\quad y = \frac{a-b+c}{a+b+c},\quad z = \frac{a+b-c}{a+b+c},$$

yielding $(b+c)/(a+b+c) = (1+x)/2$ etc., takes the claim (1) to the form

$$2 < (1+x)(1+y)(1+z) \le \frac{64}{27} \tag{2}$$

for x, y, z positive numbers with $x+y+z=1$.

Multiplying out the product in (2) we get 1, plus $x+y+z$, which is also 1, plus other positive terms; hence the left inequality of (2). The right one follows from the AM-GM inequality:

$$\begin{aligned}(1+x)(1+y)(1+z) &\le \left(\frac{(1+x)+(1+y)+(1+z)}{3}\right)^3\\ &= \left(\frac{4}{3}\right)^3 = \frac{64}{27}.\end{aligned}$$

1991/2

The numbers a_is form an arithmetic progression. Denote its step by d. Obviously $a_1 = 1$. If $d = 1$, then all positive integers less than n appear in this progression; this means that n is a prime.

Assume $d \geq 2$ for the sequel. Then the number a_2 must be a prime; otherwise it would have a prime factor p with $a_1 < p < a_2$, not appearing in the progression, hence a divisor of n, contrary to a_2 being relatively prime to n. Thus, indeed, $a_2 \geq 3$ is a prime, hence an odd number.

The step $d = a_2 - 1$ is an even number. The last term in the sequence $(a_1, \ldots, a_k)$ is $n - 1$; so we have the equality $n - 1 = a_k = 1 + (k-1)d$. Since d is even, we see that also n is even: $n = 2m$.

The disposition of terms in $(a_1, \ldots, a_k)$ is symmetric with respect to m; the number m is not coprime to n, so it does not enter the sequence. Thus the a_is split into two equally numerous groups, and therefore k is even: $k = 2j$. The two middle terms are a_j and a_{j+1}, and m lies midway between them: $a_{j+1} - m = m - a_j = d/2$.

Assume $d \geq 6$. Then the numbers $m-2$, $m-1$, $m+1$, $m+2$ lie in the interval (a_j, a_{j+1}); hence they are different from all the a_is, hence cannot be coprime to n. So the numbers n and $m - 1$ have a common prime divisor p and, likewise, the numbers n and $m - 2$ have a common prime divisor p'. Then p is also a divisor of the difference $n - 2(m-1) = 2$, and p' is a divisor of $n - 2(m-2) = 4$, implying $p = p' = 2$; a contradiction, as p and p' are divisors of the consecutive integers $m - 1$ and $m - 2$. The assumption $d \geq 6$ turned out to be wrong; so $d = 2$ or $d = 4$.

If $d = 4$, then $a_2 = 5$, $a_3 = 9$ (here the condition $n > 6$ is essential, implying that $a_2 = 5$ is not the last term in the sequence). We are again led to a contradiction: the prime number 3, lying between a_1 and a_2, ought to be a divisor of n, whereas n is coprime to $a_3 = 9$.

We are left with the case $d = 2$. Then all odd integers from 1 to $n - 1$ occur as a_is, i.e., are relatively prime to n. Consequently n is an integral power of 2. The proof is complete.

1991/3

We will call a positive integer $n \leq 280$ *nice* if it satisfies the imposed condition: each n-element subset of the set $S = \{1, 2, \ldots, 280\}$ contains five numbers which are pairwise relatively prime.

The set S contains 216 numbers that are divisible by at least one of 2, 3, 5, 7 (see the Remark). Thus if $n \leq 216$, we can find an n-element subset of S consisting entirely of integers divisible by these four initial primes; and it is impossible to choose from that subset five numbers, mutually coprime. This shows that every nice number must exceed 216. We are going to show that $n = 217$ is nice.

The set S contains 220 composite integers (see the Remark). They constitute a 220-element set, which we denote by Y. Let T be any 217-element subset of S. We need to find in the set T five numbers, pairwise relatively prime. Obviously, all elements of the set $T \setminus Y$ are pairwise coprime. Thus if this set has at least five elements, there is nothing more to do.

Now assume that the set $T \setminus Y$ has at most four elements. Then the set $Y \setminus T$ has at most seven elements. Consider the following five-element subsets of Y:

$$\begin{aligned}
M_1 &= \{2\cdot 23, & 3\cdot 19, && 5\cdot 17, && 7\cdot 13, && 11\cdot 11\},\\
M_2 &= \{2\cdot 29, & 3\cdot 23, && 5\cdot 19, && 7\cdot 17, && 11\cdot 13\},\\
M_3 &= \{2\cdot 31, & 3\cdot 29, && 5\cdot 23, && 7\cdot 19, && 11\cdot 17\},\\
M_4 &= \{2\cdot 37, & 3\cdot 31, && 5\cdot 29, && 7\cdot 23, && 11\cdot 19\},\\
M_5 &= \{2\cdot 41, & 3\cdot 37, && 5\cdot 31, && 7\cdot 29, && 11\cdot 23\},\\
M_6 &= \{2\cdot 43, & 3\cdot 41, && 5\cdot 37, && 7\cdot 31, && 13\cdot 17\},\\
M_7 &= \{2\cdot 47, & 3\cdot 43, && 5\cdot 41, && 7\cdot 37, && 13\cdot 19\},\\
M_8 &= \{2\cdot 2, & 3\cdot 3, && 5\cdot 5, && 7\cdot 7, && 13\cdot 13\}.
\end{aligned}$$

Each one of these eight (pairwise disjoint) sets consists of five pairwise coprime integers. The elements of the set $Y \setminus T$ can enter seven sets M_i (at most). There will remain at least one set M_{i_0} disjoint from $Y \setminus T$, i.e., contained in T. This gives us the five numbers, pairwise relatively prime, we were looking for. Consequently, 217 is a nice integer; in view of a previous observation, it is the smallest nice integer.

Remark. We have employed the fact that the set $S = \{1, 2, \ldots, 280\}$ contains 216 integers divisible by 2, 3, 5 or 7, and that it contains 220 composite integers. The number 280 is small enough in order that these statements do not require any justification; one can simply check that they are true. To say so, is, however, a flaw on the elegance. We now show how to avoid calculations needed for a direct verification.

For any positive integer m let S_m denote the set of all those numbers in S that are divisible by m. The following equality holds:

$$\begin{aligned}
|S_2 \cup S_3 \cup S_5 \cup S_7| = {} & (|S_2| + |S_3| + |S_5| + |S_7|) - (|S_2 \cap S_3| + |S_2 \cap S_5| \\
& + |S_2 \cap S_7| + |S_3 \cap S_5| + |S_3 \cap S_7| + |S_5 \cap S_7|) \\
& + (|S_2 \cap S_3 \cap S_5| + |S_2 \cap S_3 \cap S_7| + |S_2 \cap S_5 \cap S_7| \\
& + |S_3 \cap S_5 \cap S_7|) - |S_2 \cap S_3 \cap S_5 \cap S_7|
\end{aligned}$$

(the principle of inclusion-exclusion; the symbol $|X|$ denotes the number of elements of a finite set X). If i, j, k, l are distinct primes, then

$$S_i \cap S_j = S_{ij}, \quad S_i \cap S_j \cap S_k = S_{ijk}, \quad S_i \cap S_j \cap S_k \cap S_l = S_{ijkl}.$$

Note that $|S_m| = \lfloor 280/m \rfloor$. Therefore

$$\begin{aligned} |S_2 \cup S_3 \cup S_5 \cup S_7| &= (|S_2| + |S_3| + |S_5| + |S_7|) \\ &\quad - (|S_6| + |S_{10}| + |S_{14}| + |S_{15}| + |S_{21}| + |S_{35}|) \\ &\quad + (|S_{30}| + |S_{42}| + |S_{70}| + |S_{105}|) - |S_{210}| \\ &= (140 + 93 + 56 + 40) \\ &\quad - (46 + 28 + 20 + 18 + 13 + 8) \\ &\quad + (9 + 6 + 4 + 2) - 1 \\ &= 216, \end{aligned}$$

justifying the first one of the two statements in question.

The second one asserts that the set Y of all composite integers in S contains 220 elements. To justify that, denote the set $S_2 \cup S_3 \cup S_5 \cup S_7$ by U and note that by removing from U the four initial primes (2, 3, 5 and 7) we obtain a 212-element set of composite numbers. We have to adjoin to it all the composites not exceeding 280 and not divisible by 2, 3, 5 or 7.

Every such number has a prime factor not greater than $\sqrt{280}$, hence not exceeding 13. Thus, in order to determine the set $Y \setminus U$, we only need to examine multiples of 11 and 13, with no smaller prime factors. There are eight such numbers in S:

$$\begin{array}{llll} 11 \cdot 11 = 121, & 11 \cdot 13 = 143, & 11 \cdot 17 = 187, & 11 \cdot 19 = 209, \\ 11 \cdot 23 = 253, & 13 \cdot 13 = 169, & 13 \cdot 17 = 221, & 13 \cdot 19 = 247. \end{array}$$

Therefore $Y = (U \setminus \{2, 3, 5, 7\}) \cup \{121, 143, 169, 187, 209, 221, 247, 253\}$, so that, finally, $|Y| = 216 - 4 + 8 = 220$.

1991/4

Pick arbitrarily a vertex v_0 and start a walk from v_0, moving along the edges of G. At each junction (vertex of G) enter any edge you have not traversed yet. Label the edges by the consecutive integers $1, 2, 3, \ldots$, in the order as they appear on your route. Continue in that manner as long as it is possible. As there are only finitely many edges in G, sooner or later you will reach a vertex with no "new" edge emanating from it. Suppose the last edge on your route received label r.

If you have visited all the edges of G, end the walk. If not, there must exist at least one non-traversed edge, attached at a vertex you have already passed through; this follows from the connectedness of the graph. Choose any such edge, jump to that vertex, enter that edge, label it $r+1$ and continue the walk, according to the previous rules; label the edges on your route consecutively. As before, when you arrive at a deadlock, check whether you have passed through all the edges of G; if yes, end the walk.

If not, find a vertex on your previous trajectory, issuing at least one non-traversed edge (by connectedness, such a vertex and edge must exist). Jump to that vertex, enter that edge, and continue. Repeat that procedure as long as there are any edges left. Label the edges on your trajectory with consecutive integers. When there are no more edges in G, stop.

Each time you get stuck, you must have reached a vertex either adjacent to exactly one edge (the one that has brought you there; call it e) or a vertex issuing some other edges visited already, hence bearing labels smaller than e. This means that you must have passed through that vertex earlier.

We claim that the labeling thus obtained satisfies the required condition. The first r edges constitute the first "section" of the route. The next few traversed edges, ending at the next deadlock vertex, make up the second section. Subsequent sections are defined analogously. Note that the starting point of every section, except the first one, appears *inside* some former section; the word *inside* means that it is not an endpoint of that previous section (if it were, there would have been no obstacle to the continuation of the walk).

The assertion of the problem is obvious for vertex v_0 (belonging to the edge given label 1).

Take any vertex v of G, other than v_0, with at least two edges emanating from it. Since all edges have been traversed, this vertex must have appeared on our route. Consider its earliest appearance. As observed, it could not have appeared for the first time as a "deadlock" point, ending a section. Thus v first appeared as a point inside a certain section. So it was first reached and left through two edges of the same section, hence labeled with two consecutive integers. This is enough to conclude that the greatest common divisor of the integers labeling all the edges emanating from v is equal to 1. Since v is arbitrary, the result follows.

1991/5

First Solution. Let $\angle CAB = \alpha$, $\angle ABC = \beta$, $\angle BCA = \gamma$, $\angle PAB = \alpha'$, $\angle PBC = \beta'$, $\angle PCA = \gamma'$. We may assume that $\alpha \geq 30°$, $\beta \geq 30°$, $\gamma \geq 30°$; otherwise the claim is obvious. Applying the law of sines to the triangles PAB, PBC, PCA we see that

$$\frac{PA}{PB} = \frac{\sin(\beta - \beta')}{\sin\alpha'}, \qquad \frac{PB}{PC} = \frac{\sin(\gamma - \gamma')}{\sin\beta'}, \qquad \frac{PC}{PA} = \frac{\sin(\alpha - \alpha')}{\sin\gamma'}.$$

The product of these three numbers is equal to 1 by Ceva's theorem; see *Glossary*. Writing

$$f(\theta) = \frac{\sin(\alpha - \theta)}{\sin\theta}, \qquad g(\theta) = \frac{\sin(\beta - \theta)}{\sin\theta}, \qquad h(\theta) = \frac{\sin(\gamma - \theta)}{\sin\theta}$$

we obtain the equality $f(\alpha')g(\beta')h(\gamma') = 1$. Hence, by the AM-GM inequality,

$$f(\alpha') + g(\beta') + h(\gamma') \geq 3. \tag{1}$$

Note that f is a decreasing function of the variable $\theta \in (0, 180°)$; this is apparent from $f(\theta) = \sin\alpha\cot\theta - \cos\alpha$. Similarly, g and h are decreasing functions. For any angles φ, $\psi \in (0, 180°)$ we have

$$\sin\varphi + \sin\psi = 2\sin\frac{\varphi + \psi}{2}\cos\frac{\varphi - \psi}{2} \leq 2\sin\frac{\varphi + \psi}{2}. \tag{2}$$

Set in (2) $\varphi = \alpha - 30°$, $\psi = \beta - 30°$, and then $\varphi = \gamma - 30°$, $\psi = 30°$ (these are nonnegative angle measures, according to the assumption made at the beginning):

$$\sin(\alpha - 30°) + \sin(\beta - 30°) \leq 2\sin\frac{\alpha + \beta - 60°}{2},$$

$$\sin(\gamma - 30°) + \frac{1}{2} \leq 2\sin\frac{\gamma}{2}.$$

Adding these inequalities and applying to the sum that arises on the right the inequality (2) with $\varphi = (\alpha + \beta - 60°)/2$, $\psi = \gamma/2$, we hence obtain

$$\begin{aligned} &\sin(\alpha - 30°) + \sin(\beta - 30°) + \sin(\gamma - 30°) + \frac{1}{2} \\ &\leq 4\sin\frac{\alpha + \beta + \gamma - 60°}{4} \\ &= 2. \end{aligned}$$

In view of $f(30°) = 2\sin(\alpha - 30°)$ and analogous expressions for the values $g(30°)$ and $h(30°)$, we arrive at the inequality

$$f(30^\circ) + g(30^\circ) + h(30^\circ) \le 3. \tag{3}$$

(This can be also deduced from Jensen's inequality for the sine function, concave in $(0, 180^\circ)$; see *Glossary*.) Inequality (3), combined with (1), shows that

$$f(\alpha') \ge f(30^\circ) \quad \text{or} \quad g(\beta') \ge g(30^\circ) \quad \text{or} \quad h(\gamma') \ge h(30^\circ).$$

And since the functions f, g, h are decreasing, we conclude that at least one of the angles α', β', γ' is not greater than 30°. This is precisely the assertion of the problem.

Second Solution. No trigonometry, no calculations, pure geometry. We prove a stronger result:

Proposition. *Inside every triangle ABC there is a point Q such that*

$$\angle QAB = \angle QBC = \angle QCA \le 30^\circ.$$

This granted, the solution of the problem is automatic: let ABC be the given triangle and let Q be as above; it suffices to note that the given point P belongs to one of the triangles QAB, QBC, QCA, and the claim is plain. So it remains to prove the proposition.

Proof of the proposition. Assume without loss of generality that C is the greatest angle of triangle ABC. Define Q to be the point of intersection (other than A) of the following two circles:

circle ω, tangent to AB at A, passing through C;

circle ω', tangent to BC at B, passing through A.

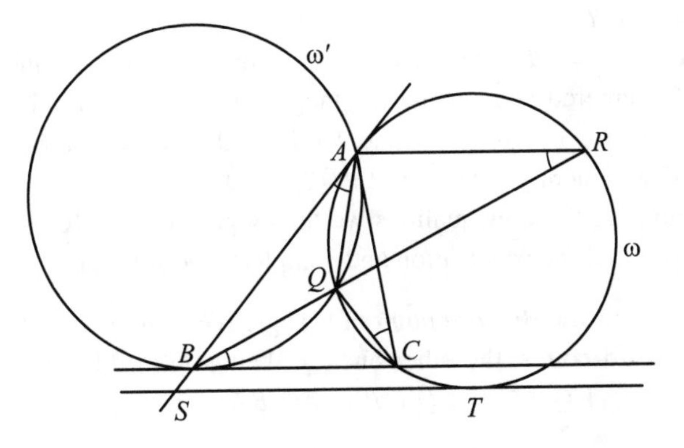

(They are not tangent to each other at A, hence they intersect at two distinct points, A and Q.) Point B lies outside circle ω, and C lies outside ω'. Consequently, Q is an interior point of triangle ABC.

From the tangent-chord theorem, applied to circles ω and ω', we see that $\angle QCA = \angle QAB$ (circle ω) and $\angle QBC = \angle QAB$ (circle ω'), and we have the part of the claim asserting that the angles QAB, QBC, QCA are equal (see the Remark). We still have to show that their common size does not exceed $30°$.

Ray BQ cuts ω at a second point, which we denote by R. Since $\angle QRA = \angle QCA = \angle QBC$, lines AR and BC are parallel. Draw the line parallel to these two, tangent to ω at T (on the same side of AR as C); it cuts line AB at a point S. Thus

$$\angle QBC = \angle RBC \leq \angle RST.$$

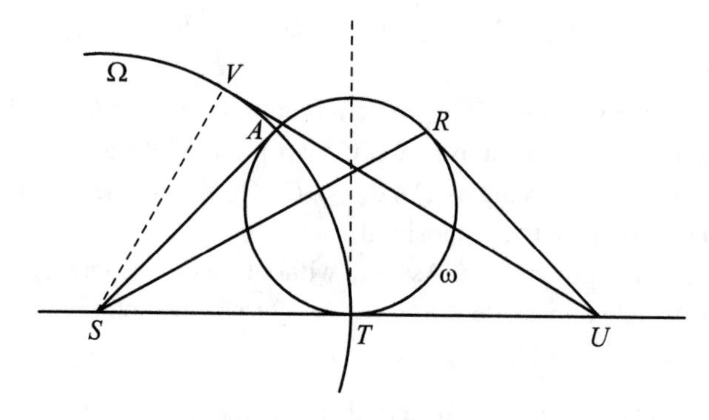

Let U be the point symmetric to S with respect to T. The diameter of ω through T is the axis of symmetry of the trapezoid $ASUR$. Hence $\angle RST = \angle AUT$.

Since $SA = ST$ (the tangents to ω from S), points A and T lie on a circle Ω centered at S. Draw the tangent UV to Ω, with V on Ω (on the same side of line SU as A and R). Note that SUV is a $60°$–$30°$–$90°$ triangle. Consequently $\angle AUT \leq \angle VUT = 30°$.

Combining these inequalities we finally get that $\angle QBC \leq 30°$. This ends the proof of the proposition and completes the solution.

Remark. Q is the *Brocard point* of triangle ABC; to be precise, it is one of two Brocard points, the other one, Q', being defined by the analogous angle equalities $\angle Q'AC = \angle Q'CB = \angle Q'BA$.

1991/6

First Solution. A sequence with properties as required can be constructed even for $a = 1$. The reasoning will be based on the following simple fact about approximation of $\sqrt{2}$ by rational numbers:

$$\left|\sqrt{2} - \frac{m}{n}\right| \geq \frac{1}{n(m + \sqrt{2}n)} \tag{1}$$

for each pair of integers $m, n \geq 1$. To justify this, note that $|2n^2 - m^2| \geq 1$ (because $\sqrt{2}$ is irrational); hence

$$\left|\sqrt{2} - \frac{m}{n}\right| = \frac{|\sqrt{2}n - m|}{n} = \frac{|2n^2 - m^2|}{n(\sqrt{2}n + m)} \geq \frac{1}{n(m + \sqrt{2}n)},$$

proving (1). Fix integers $i > j \geq 0$ and set

$$n = i - j, \qquad m = \lfloor\sqrt{2}i\rfloor - \lfloor\sqrt{2}j\rfloor. \tag{2}$$

Since $m < \sqrt{2}i - \sqrt{2}j + 1 = \sqrt{2}n + 1 \leq (\sqrt{2} + 1)n$, we see that

$$m + \sqrt{2}n < Cn, \qquad \text{where} \quad C = 2\sqrt{2} + 1.$$

From (1) and (2) we get

$$\left|\sqrt{2} - \frac{\lfloor\sqrt{2}i\rfloor - \lfloor\sqrt{2}j\rfloor}{i - j}\right| \geq \frac{1}{C(i - j)^2};$$

equivalently:

$$C \cdot \left|\sqrt{2}(i - j) - \left(\lfloor\sqrt{2}i\rfloor - \lfloor\sqrt{2}j\rfloor\right)\right| \cdot (i - j) \geq 1. \tag{3}$$

It is now enough to define

$$x_i = C\left(\sqrt{2}i - \lfloor\sqrt{2}i\rfloor\right) \qquad \text{for } i = 0, 1, 2, \ldots.$$

Evidently, $|x_i| \leq C$ for all i; the sequence $x_0, x_1, x_2, \ldots$ is bounded. And according to inequality (3), $|x_i - x_j| \cdot (i - j) \geq 1$ for $i > j \geq 0$, which automatically implies the weaker claim

$$|x_i - x_j| \cdot (i - j)^a \geq 1 \quad \text{for } i > j \geq 0,$$

for any real number $a \geq 1$.

Second Solution. In this solution the condition $a > 1$ will be used; the construction of the sequence $x_0, x_1, x_2, \ldots$ will essentially depend on the parameter a.

Fix a real number $a > 1$. It is a well known fact that there exists a finite constant C_a such that

$$\frac{1}{1^a} + \frac{1}{2^a} + \cdots + \frac{1}{n^a} < C_a \quad \text{for } n = 1, 2, 3, \ldots. \tag{4}$$

(In other words, the infinite series $1 + (\frac{1}{2})^a + (\frac{1}{3})^a + \cdots$ converges; a proof can be found in any textbook on calculus.)

The definition of the sequence $x_0, x_1, x_2, \ldots$ will be inductive. Let $x_0 = 0$. Fix an integer $k \geq 1$ and assume we have already defined real numbers $x_0, \ldots, x_{k-1}$ satisfying the conditions

$$|x_i| \leq C_a, \qquad |x_i - x_j| \cdot |i - j|^a \geq 1 \tag{5}$$

for $i, j \in \{0, \ldots, k-1\}$, $i \neq j$. Consider the closed intervals

$$J_i = [x_i - \delta_i, x_i + \delta_i] \quad \text{where } \delta_i = \frac{1}{(k-i)^a}, \qquad i = 0, \ldots, k-1.$$

Their joint length does not exceed $2\delta_0 + \cdots + 2\delta_{k-1}$; this sum is less than $2C_a$, in accordance with (4). So we can find a number $x_k \in [-C_a, C_a]$ that does not belong to any one of the intervals $J_0, \ldots, J_{k-1}$. This means that the following inequalities hold for $i = 0, \ldots, k-1$:

$$|x_k| \leq C_a, \qquad |x_k - x_i| > \delta_i = \frac{1}{(k-i)^a},$$

showing that conditions (5) are satisfied for $i, j \in \{0, \ldots, k\}$, $i \neq j$, and completing the induction step.

As a result of this inductive procedure we obtain an infinite sequence $x_0, x_1, x_2, \ldots$ whose terms satisfy conditions (5) for every pair of distinct nonnegative integers i, j.

Thirty-third International Olympiad, 1992

1992/1

Let a, b, c be numbers satisfying the given conditions and let $a = x + 1$, $b = y + 1$, $c = z + 1$. Then x, y, z are integers with $1 \leq x < y < z$ and their product xyz is a divisor of $abc - 1$. Thus the number

$$\frac{(x+1)(y+1)(z+1) - 1}{xyz} = \frac{xyz + yz + zx + xy + x + y + z}{xyz}$$

is an integer. Denote it by $k+1$. Then

$$\frac{1}{x}+\frac{1}{y}+\frac{1}{z}+\frac{1}{yz}+\frac{1}{zx}+\frac{1}{xy}=k. \tag{1}$$

Since $1 \le x < y < z$, we see that $y \ge 2$, $z \ge 3$, and equation (1) gives the estimate $0 < k \le 17/6$; so k is either 1 or 2.

Assume that $x \ge 3$. Then $y \ge 4$, $z \ge 5$, and we obtain from (1) $k \le 59/60$, a contradiction. Thus also x equals either 1 or 2.

If $x = 1$, then the first term in (1) is 1; so $k = 2$ and (1) after recasting becomes $(y-2)(z-2) = 5$. In view of $2 \le y < z$ we get $y = 3$, $z = 7$.

The other possible case is $x = 2$; equation (1) takes the form

$$\frac{1}{2}+\frac{3}{2}\left(\frac{1}{y}+\frac{1}{z}\right)+\frac{1}{yz}=k. \tag{2}$$

Now $y \ge 3$, $z \ge 4$, whence by (2) $k \le 35/24$; i.e., $k = 1$, and (2) easily becomes $(y-3)(z-3) = 11$. From the inequalities $3 \le y < z$ we now get $y = 4$, $z = 14$.

The two triples $(x, y, z) = (1, 3, 7)$ and $(x, y, z) = (2, 4, 14)$ yield the triples $(a, b, c) = (2, 4, 8)$ and $(a, b, c) = (3, 5, 15)$. Each of them satisfies the conditions of the problem.

1992/2

Suppose a function $f\colon \mathrm{R} \to \mathrm{R}$ satisfies the given equation

$$f(x^2 + f(y)) = y + (f(x))^2 \quad \text{for } x, y \in \mathrm{R}. \tag{1}$$

Denote $f(0) = c$, $f(1) = d$. Setting in (1) $x = 0$ we get

$$f(f(y)) = y + c^2 \quad \text{for } y \in \mathrm{R}. \tag{2}$$

The last equation with $y = 0$ and $y = 1$ gives $f(c) = c^2$ and $f(d) = 1+c^2$.

Apply the function f to both sides of (1) to obtain

$$f(f(x^2 + f(y))) = f(y + (f(x))^2) \quad \text{for } x, y \in \mathrm{R}.$$

This in view of (2) is equivalent to

$$x^2 + f(y) + c^2 = f(y + (f(x))^2) \quad \text{for } x, y \in \mathrm{R}. \tag{3}$$

Setting in (3) $x = 1$, $y = c$, and remembering that $f(c) = c^2$, we get $1 + 2c^2 = f(c + d^2)$; while setting in (1) $x = d$, $y = 0$, and remembering that $f(d) = 1 + c^2$, we get $f(d^2 + c) = (1 + c^2)^2$. These equalities imply

$1 + 2c^2 = (1 + c^2)^2$, whence $c = 0$. Equation (2) becomes

$$f(f(x)) = x \quad \text{for } x \in \mathrm{R}. \tag{4}$$

We claim that the function f is strictly increasing. Let s and t be any real numbers such that $s < t$ and let $u = f(\sqrt{t-s})$. By (4), $f(u) = \sqrt{t-s} > 0$ (hence $u \neq 0$). Applying equation (3) with $x = u$, $y = s$, and keeping in mind that $c = 0$, we obtain $u^2 + f(s) = f(t)$. Thus $f(s) < f(t)$, showing that f is strictly increasing.

Now it is easy to show that

$$f(x) = x \quad \text{for all } x \in \mathrm{R}.$$

Indeed; assuming $f(v) > v$ for a certain $v \in \mathrm{R}$, we get from (4) and by the monotonicity of f: $v = f(f(v)) > f(v)$, contrary to hypothesis. Also assuming $f(v) < v$ for some $v \in \mathrm{R}$ we are analogously led to a contradiction.

Thus f must be the identity function $f(x) = x$. Evidently, this function satisfies equation (1). So it is the unique solution.

1992/3

Label the points 1 through 9 and color blue every edge such that the labels of its endpoints differ by 1, 4 or 6. We claim that no three blue segments form a triangle. To see this, consider a triangle spanned by vertices labeled i, j, k $(i > j > k)$. If all its sides were blue, the differences $i - j = x$, $j - k = y$ and $i - k = x + y$ would belong to the set $\{1, 4, 6\}$. However, no element of this set is equal to the sum of any two other elements (distinct or not); we say that the set $\{1, 4, 6\}$ is *sum-free*. This settles our claim.

Now consider the set $\{2, 3, 7, 8\}$, which is also sum-free. If we color red every edge whose endpoints bear labels differing by 2, 3, 7 or 8, then, analogously, no red triangle shall appear.

Only those edges have been left uncolored that connect pairs of points whose labels differ by 5. There are four such pairs (i, j) $(i > j)$, namely, $(6, 1)$, $(7, 2)$, $(8, 3)$, $(9, 4)$. The total number of edges is $\binom{9}{2} = 36$. So we have shown how to color 32 edges in two colors so that no *monochromatic* triangle should arise.

Now we show that this number cannot be increased. Suppose we have colored 33 segments. Choose one endpoint of each one of the remaining three segments and color the chosen endpoints brown; some of these brown points can coincide, but we need not care about that. Color yellow the other points (out of the given nine); there are at least six of them. Every uncolored segment has a brown endpoint, and therefore all the edges connecting yellow points are colored, either blue or red.

Take any yellow vertex ℓ. Colored edges link it to other yellow points, no fewer than five. Thus at least three of these edges have the same color, say, red. Let i, j, k be their endpoints other than ℓ. They are yellow, so the three edges which pairwise connect them have been colored. If one of these three edges is red, then its endpoints together with point ℓ span a red triangle. If not, then these three edges are blue and we have a blue triangle. Anyway, a monochromatic triangle unavoidably appears.

Concluding, $n = 33$ is the least integer such that whenever n edges are colored in the manner considered, a monochromatic triangle must appear.

Remark. Consider a finite set of points together with the family of all edges connecting pairs of those points; such a configuration is called a *complete graph* (there is no "geometry" in this concept, just set theory or combinatorics; see the statement of problem 1991/4). If we pick from such a graph some points together with all edges determined by these points, we obtain a smaller complete graph, which is often called a *clique* (in the original graph). The following theorem is true:

Let k and m be positive integers; there exists an integer n such that, for every two-coloring of all the edges in a complete n-point graph, there necessarily appears either a k-point clique with all edges of the first color or an m-point clique with all edges of the second color.

This is *Ramsey's theorem*; or, to be more precise, one of its specific cases. (In a more general setting, Ramsey's theorem involves coloring in more than two colors; yet more general variants can be considered.)

The smallest n satisfying the assertion of the above theorem, for given integers k and m, is called the *Ramsey number* of the pair (k, m) and is denoted by $R(k, m)$. In the final section of the solution of the problem (dealing with six yellow points) we have in fact shown that $R(3, 3) = 6$.

1992/4

First Solution. Let line L touch C at T. Suppose P is a point with the property in question; thus there are points Q and R on L satisfying the given condition. We may assume that the points Q, T, M, R lie on line L in that order (except that T and M can coincide). Let TV be the diameter of C. Let S be the foot of the perpendicular to L from P and let U complete the rectangle $STVU$. Denote by I and r the center and radius of C. Finally, denote: $QR = a$, $RP = b$, $PQ = c$, $PS = h$; then $b \geq c$; points M and S lie on L on distinct sides of T (or they both coincide with T). Since T is the point of contact of the incircle of triangle PQR with side QR,

$$TM = QM - QT = \frac{a}{2} - \frac{a-b+c}{2} = \frac{b-c}{2}. \tag{1}$$

Using the equality $ah = (a+b+c)r$ $(= 2\text{area}(\Delta PQR))$ we get

$$PU = h - 2r = \frac{ah - 2ar}{a} = \frac{(b+c-a)r}{a}. \tag{2}$$

If the angle PQR is nonobtuse, as in the diagram, then

$$SR + SQ = a \quad \text{and} \quad SR - SQ = 2 \cdot SM;$$

and if the angle PQR is obtuse then, conversely,

$$SR + SQ = 2 \cdot SM \quad \text{and} \quad SR - SQ = a.$$

In either case, $2a \cdot SM = SR^2 - SQ^2 = b^2 - c^2$. Hence by (1)

$$UV = SM - TM = \frac{b^2 - c^2}{2a} - \frac{b-c}{2} = \frac{(b-c)(b+c-a)}{2a}. \tag{3}$$

Dividing (3) by (2) and using (1) again we obtain

$$\frac{UV}{PU} = \frac{b-c}{2r} = \frac{TM}{IT}.$$

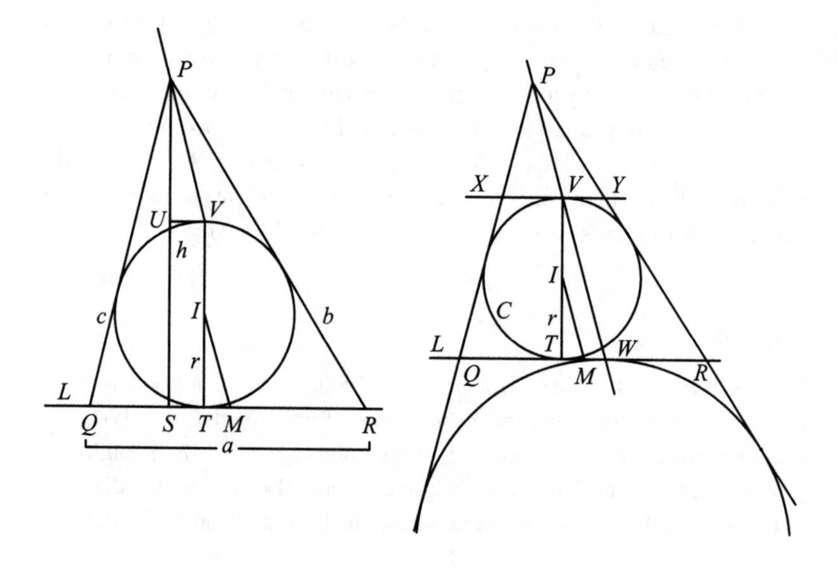

Thus the right triangles PUV and ITM are similar. And since M and P lie on distinct sides of line IT, we infer that $VP \parallel IM$. Consequently, P lies on the ray parallel to IM originating at V and not intersecting L.

This ray (without its origin V) is in fact the locus we seek; to justify that, pick any point P on that ray (not V) and draw the two tangents from P to circle C, intercepting a segment QR on L. Denote its midpoint by M'. According to the first part of the proof, $VP \parallel IM'$; and from the definition of the ray VP it follows that $VP \parallel IM$. This means that M coincides with M'; i.e., it is the midpoint of QR. So the point P has the property in question; the ray defined above is the locus sought.

Second Solution. This is, in fact, a variation of the previous method. Let the points T, I, V be defined as above. Assume P is a point with the property considered and let Q, R be points on L, as specified in the problem statement. Draw the other tangent to C, parallel to L. It passes through V and cuts lines PQ, PR in the respective points X, Y. Line PV cuts L at a point W.

Consider the homothety of ratio PW/PV, centered at P; it carries triangle PXY onto PQR. Since C is the circle escribed to triangle PXY, tangent to XY at V, its image is the circle escribed to triangle PQR, tangent to QR at W.

In triangle PQR, the points T and W, at which the side QR touches the incircle and the excircle, are situated symmetrically with respect to M, the midpoint of that side. This is a well-known fact (if not, it is proposed as an easy exercise).

Thus IM is the segment connecting the midpoints of sides TV and TW of triangle VTW. Consequently, it is parallel to side VW, i.e., to line PV. We hence conclude that the point P lies on the ray defined in the first solution (parallel to IM, originating at V and not intersecting L).

Moreover, every point of that ray, except its origin V, is a feasible position of P. To justify that, we repeat the last section of the previous solution.

1992/5

Visualize the xy-plane horizontally and z-axis vertically. Let $z_1 < \cdots < z_n$ be all the values taken by the z-coordinate of points of S. The whole set S is sliced into "layers" $L_1, \ldots, L_n$; layer L_i is defined as the horizontal section of S by the plane $z = z_i$. Assume that the x-coordinates of the points in L_i take a_i distinct values and the y-coordinates of those points take b_i distinct values. In other words, a_i and b_i are the cardinalities of the orthogonal projections of L_i onto the xz-plane and yz-plane, respectively. Thus

$$|S_y| = a_1 + \cdots + a_n \quad \text{and} \quad |S_x| = b_1 + \cdots + b_n. \tag{1}$$

Layer L_i consists of certain points of intersection of the a_i lines parallel to the y-axis and the b_i lines parallel to the x-axis. Therefore $|L_i| \le a_i b_i$. Projecting L_i onto the xy-plane we obtain a subset of S_z, so that $|L_i| \le |S_z|$. Denote $|S_z|$ by c; the estimate follows: $|L_i|^2 \le a_i b_i c$. Consequently,

$$|S| = |L_1| + \cdots + |L_n| \le \left(\sqrt{a_1 b_1} + \cdots + \sqrt{a_n b_n}\right) \cdot \sqrt{c}. \tag{2}$$

The sum in the parentheses can be estimated using the Cauchy–Schwarz inequality (see *Glossary*):

$$\left(\sqrt{a_1 b_1} + \cdots + \sqrt{a_n b_n}\right)^2 \le (a_1 + \cdots + a_n)(b_1 + \cdots + b_n). \tag{3}$$

From (2), (3) and (1) we immediately obtain the claimed inequality:

$$|S|^2 \le (a_1 + \cdots + a_n)(b_1 + \cdots + b_n) \cdot c = |S_x| \cdot |S_y| \cdot |S_z|.$$

1992/6

(a) Suppose that the square of a certain integer $n \ge 4$ has been written as the sum of k positive square integers:

$$n^2 = m_1^2 + \cdots + m_k^2 \qquad (m_1, \ldots, m_k \ge 1).$$

Denote by r the greatest number among $m_1, \ldots, m_k$; assume that among those numbers there are i_1 "ones," i_2 "twos," i_3 "threes," etc., i_r terms equal to r ($i_1 \ge 0, i_2 \ge 0, \ldots, i_{r-1} \ge 0, i_r \ge 1$). Thus

$$i_1 + 4i_2 + 9i_3 + \cdots + r^2 i_r = n^2, \quad i_1 + i_2 + i_3 + \cdots + i_r = k.$$

Subtracting the second equality from the first one,

$$3i_2 + 8i_3 + \cdots + (r^2 - 1)i_r = n^2 - k.$$

If $r \ge 4$, then by the last equality $n^2 - k \ge 15$, so that $k < n^2 - 13$. And if $r \le 3$, we get $3i_2 + 8i_3 = n^2 - k$; the sum $3i_2 + 8i_3$ does not assume value 13 (for any integers $i_2, i_3 \ge 0$). Consequently, k never can be equal to $n^2 - 13$. This automatically implies $S(n) \le n^2 - 14$.

(b) In order that equality $S(n) = n^2 - 14$ be satisfied, the number n^2 has to be representable (in particular) as the sum of two squares and as the sum of three squares. This suggests a glance at Pythagorean triples: $(3, 4, 5)$, $(6, 8, 10)$, $(5, 12, 13)$, ...

It is easy to verify that neither $5^2 = 25$ nor $10^2 = 100$ can be written as the sum of three squares. Trials with the number $13^2 = 169$ show that it can be written as the sum of k squares, for small values of k:

$$\begin{aligned}169 &= 144 + 25 = 144 + 16 + 9 = 121 + 16 + 16 + 16\\ &= 36 + 36 + 36 + 36 + 25.\end{aligned}$$

This gives some hope that 169 might be "good." The following lemma will be the technical tool, useful in the solution of part (b) and (c):

Lemma. *If integers N and k satisfy the inequalities $15 \le N \le 4k$, $k \le N - 14$, then N can be written as the sum of the squares of k numbers, each of which is equal to 1 or 2 or 3.*

Proof of the lemma. The claim is that the system of equations

$$x + y + z = k, \qquad 9x + 4y + z = N \tag{1}$$

has a solution in nonnegative integers x, y, z. In fact, it suffices to take, for example,

$$x = \begin{cases} 2 & \text{if } N - k \equiv 1 \pmod 3,\\ 1 & \text{if } N - k \equiv 2 \pmod 3,\\ 0 & \text{if } N - k \equiv 0 \pmod 3,\end{cases}$$

$$y = \frac{N - k - 8x}{3},$$

$$z = k - x - y.$$

From the definition of x it follows that y is an integer. The first equation of (1) is satisfied obviously, and the second one is verified easily:

$$9x + 4y + z = 9x + 4y + (k - x - y) = 8x + k + 3y = N.$$

It remains to justify that the integers x, y, z are nonnegative. We must use the inequalities from the conditions of the lemma:

—if $N - k \ge 16$, then $3y = N - k - 8x \ge 0$ (as $x \le 2$);

—if $N - k = 15$, then $x = 0$, and so $3y = N - k - 8x = 15 > 0$;

—if $N - k = 14$, then $x = 1$, so that $3y = N - k - 8x = 6 > 0$.

Finally,

$$z = k - x - y = k - x - \frac{N - k - 8x}{3} = \frac{4k - N + 5x}{3} \geq \frac{4k - N}{3} \geq 0,$$

ending the proof of the lemma.

Apply the lemma to $N = 169$; its conditions are fulfilled for each integer k with $43 \leq k \leq 155$. This yields the required representation

$$169 = m_1^2 + \cdots + m_k^2 \qquad (m_1, \ldots, m_k \geq 1) \tag{2}$$

for $k = 43, 44, \ldots, 155$. It can be also obtained for $k = 1, 2, 3, 4, 5$, as shown at the beginning of part (b).

Taking $N = 36$ in the lemma we can write

$$36 = l_1^2 + \cdots + l_k^2 \qquad (l_1, \ldots, l_k \geq 1) \tag{3}$$

for any integer k with $9 \leq k \leq 22$. Representation (3) can be also obtained for $k = 1, 3, 4, 5, 6, 7, 8$ as follows:

$$\begin{aligned} 36 &= 16 + 16 + 4 = 9 + 9 + 9 + 9 = 16 + 9 + 9 + 1 + 1 \\ &= 16 + 4 + 4 + 4 + 4 + 4 = 16 + 9 + 4 + 4 + 1 + 1 + 1 \\ &= 9 + 9 + 4 + 4 + 4 + 4 + 1 + 1. \end{aligned}$$

Using the equality $169 = 36 + 36 + 36 + 36 + 25$, combined with the fact that 25 is the sum of two squares and that 36 can be written as the sum of any number of squares, not equal to 2 and not greater than 22, we can obtain representation (2) for each $k \in \{5, 6, \ldots, 42\}$, in many ways.

Thus representation (2) can be realized for every $k \in \{1, 2, \ldots, 155\}$. And this means that $S(13) = 13^2 - 14$.

(c) We are going to prove that

$$\text{if} \quad S(n) = n^2 - 14, \quad \text{then also} \quad S(2n) = (2n)^2 - 14. \tag{4}$$

Since the equality $S(n) = n^2 - 14$ holds for $n = 13$, implication (4) guarantees that it holds for infinitely many integers n.

For a proof of (4), consider an $n \geq 13$ satisfying $S(n) = n^2 - 14$ and apply the lemma taking $N = 4n^2$. By the assertion of the lemma, we have the representation

$$4n^2 = m_1^2 + \cdots + m_k^2 \qquad (m_1, \ldots, m_k \geq 1) \tag{5}$$

for every integer k such that $n^2 \leq k \leq 4n^2 - 14$.

The equality $S(n) = n^2 - 14$ ensures that n^2 can be written as

$$n^2 = l_1^2 + \cdots + l_k^2 \qquad (l_1, \ldots, l_k \geq 1) \tag{6}$$

for $k = 1, 2, \ldots, n^2 - 14$. Obviously, representation (6) induces representation (5): $4n^2 = (2l_1)^2 + \cdots + (2l_k)^2$. Representation (5) is therefore possible for $k \in \{1, 2, \ldots, n^2 - 14\} \cup \{n^2, n^2 + 1, \ldots, 4n^2 - 14\}$.

There remains a gap to be filled: the values of k from $n^2 - 13$ to $n^2 - 1$. We have the following remedy to this last obstacle:

$$4n^2 = 64 + \underbrace{4 + \cdots + 4}_{n^2-21} + \underbrace{4 + \cdots + 4}_{j} + \underbrace{1 + \cdots + 1}_{20-4j}, \tag{7}$$

$$4n^2 = 36 + 16 + 16 + \underbrace{4 + \cdots + 4}_{n^2-22} + \underbrace{4 + \cdots + 4}_{j} + \underbrace{1 + \cdots + 1}_{20-4j}, \tag{8}$$

$$4n^2 = 36 + 36 + \underbrace{4 + \cdots + 4}_{n^2-23} + \underbrace{4 + \cdots + 4}_{j} + \underbrace{1 + \cdots + 1}_{20-4j}, \tag{9}$$

where j can be any number from the set $\{0, 1, 2, 3, 4, 5\}$. The right-hand expressions in (7), (8) and (9) have (respectively) $n^2 - 3j$, $n^2 + 1 - 3j$ and $n^2 - 1 - 3j$ summands. The needed values of k (from $n^2 - 13$ to $n^2 - 1$) can be obtained by a suitable choice of j.

In conclusion, representation (5) can be realized for every k from 1 to $4n^2 - 14$. This completes the proof of implication (4), hence also the solution of the problem.

Thirty-fourth International Olympiad, 1993

1993/1

First Solution. Assume $f(x)$ is the product of polynomials $F(x)$ and $G(x)$,

$$f(x) = x^n + 5x^{n-1} + 3 = F(x)G(x), \tag{1}$$

with the given properties:

$$F(x) = a_0 + a_1 x + \cdots + a_k x^k, \quad G(x) = b_0 + b_1 x + \cdots + b_m x^m;$$

a_i, b_j integers; $0 < k < n$, $0 < m < n$, $k + m = n$. The coefficient of x^n in the product $F(x)G(x)$ is $a_k b_m$, and the free term is $a_0 b_0$. Thus $a_0 b_0 = 3$, $a_k b_m = 1$, so that $a_k = b_m = \pm 1$, one of the numbers a_0, b_0 equals 3 or

-3, and the other one is 1 or -1. There is no loss of generality in assuming $a_0 = \pm 1$, $b_0 = \pm 3$.

Let ℓ be the least index for which b_ℓ is not divisible by 3. Since $b_0 = \pm 3$ and $b_m = \pm 1$, we see that $0 < \ell \le m$. The coefficient of x^ℓ in $F(x)G(x)$ equals

$$a_\ell b_0 + a_{\ell-1} b_1 + \cdots + a_1 b_{\ell-1} + a_0 b_\ell \tag{2}$$

(we define $a_i = 0$ for $i > k$). The term $a_0 b_\ell$ is not divisible by 3; all the other terms in (2) are divisible by 3 (because so are the numbers b_j for $j < \ell$). Hence the sum (2) is not divisible by 3.

In the polynomial (1), x^n and x^{n-1} only occur with coefficients not divisible by 3. Therefore $\ell \ge n-1$; and since $\ell \le m < n$, we eventually get $\ell = m = n-1$. Thus $k = 1$ and $F(x) = a_0 + a_1 x$.

Recalling that $a_0 = \pm 1$ and $a_1 (= a_k) = \pm 1$, we infer that one of the numbers 1 and -1 is a root of $F(x)$, hence also of $f(x)$. However,

$$f(1) = 9, \quad f(-1) = 3 + 4 \cdot (-1)^{n-1} \neq 0 \quad \text{for all } n \in \mathrm{N}. \tag{3}$$

This proves that factorization (1) is not possible.

Second Solution. There is a standard method of handling such problems via polynomial algebra in the complex domain. Assume that the polynomial $f(x)$ admits factoring (1), with $F(x)$ and $G(x)$ as in the problem statement. Then $f(0) = 3$ is the product of the integers $F(0)$ and $G(0)$, one of which must be 3 or -3, and the other 1 or -1. As in the first solution, assume without loss of generality that $F(0) = \pm 1$, $G(0) = \pm 3$.

The polynomial $f(x)$ has n complex roots $z_1, \ldots, z_n$, counting multiplicities. Each of them is a root of either $F(x)$ or $G(x)$. Let $z_1, \ldots, z_k$ be roots of $F(x)$, and $z_{k+1}, \ldots, z_n$ be roots of $G(x)$. The degrees of these two polynomials add up to n, and hence

$$F(x) = A(x - z_1) \cdots (x - z_k), \quad G(x) = B(x - z_{k+1}) \cdots (x - z_n)$$

with nonzero integer constants A and B.

By assumption, these polynomials are nonconstant; so $0 < k < n$. The product of A and B is equal to the leading coefficient in the polynomial $F(x)G(x) = f(x)$, which is 1. Thus $A = B = 1$ or $A = B = -1$.

The numbers $F(0) = (-1)^k A z_1 \cdots z_k$, $G(0) = (-1)^{n-k} B z_{k+1} \cdots z_n$ are the free terms of the two polynomials. Recalling that $F(0) = \pm 1$ we get that

$$z_1 \cdots z_k = 1 \quad \text{or} \quad z_1 \cdots z_k = -1. \tag{4}$$

The numbers z_j (roots of $f(x)$) satisfy the equations $z_j^{n-1}(z_j+5) = -3$ which, multiplied over $j = 1, \ldots, k$, yield (in view of (4)):

$$|(z_1+5)\cdots(z_k+5)| = 3^k.$$

The expression on the left side has value $|F(-5)|$. Therefore

$$|f(-5)| = |F(-5)G(-5)| = 3^k \cdot |G(-5)|.$$

On the other hand, $f(-5) = 3$ (by the definition of $f(x)$). And since $|G(-5)|$ is an integer, we infer that 3^k is a divisor of 3. The exponent k is positive; so $k = 1$. Condition (4) gets reduced to $z_1 = \pm 1$. This yields the desired contradiction because the values $f(1)$ and $f(-1)$ are different from zero; see calculation (3) in the first solution. The result follows.

1993/2

Let K be an arbitrary point on ray CD produced beyond D. The external angles of triangles ADC and BDC satisfy the equalities $\angle ADK = \angle CAD + \angle ACD$ and $\angle BDK = \angle CBD + \angle BCD$. Adding them we get $\angle ADB = \angle CAD + \angle CBD + \angle ACB$. This, in view of the first condition of the problem, implies

$$\angle CAD + \angle CBD = 90^\circ. \tag{1}$$

This equality is the clue to the solution of both parts of the problem.

(a) On side BC of triangle ABC erect outwardly triangle CBE, directly similar to CAD. Thus

$$\angle CAD = \angle CBE, \quad \angle ACD = \angle BCE \tag{2}$$

and

$$\frac{AD}{AC} = \frac{BE}{BC}, \qquad \frac{CA}{CB} = \frac{CD}{CE}. \tag{3}$$

The second equality of (2) implies $\angle ACB = \angle DCE$; this, combined with the second equality of (3), shows that triangle ABC is similar to triangle DEC, and therefore

$$\frac{AB}{AC} = \frac{DE}{DC}. \tag{4}$$

The first equality of (2), accompanied by formula (1), gives

$$\angle DBE = \angle CBD + \angle CBE = \angle CBD + \angle CAD = 90^\circ.$$

Finally, the first equality of (3) and the second condition of the problem ($AC \cdot BD = AD \cdot BC$) jointly result in the equality $BD = BE$. Hence DBE is an isosceles right triangle, and so $DE = \sqrt{2} \cdot BD$. Inserting this into formula (4) we obtain the outcome in part (a):

$$\frac{AB \cdot CD}{AC \cdot BD} = \sqrt{2}.$$

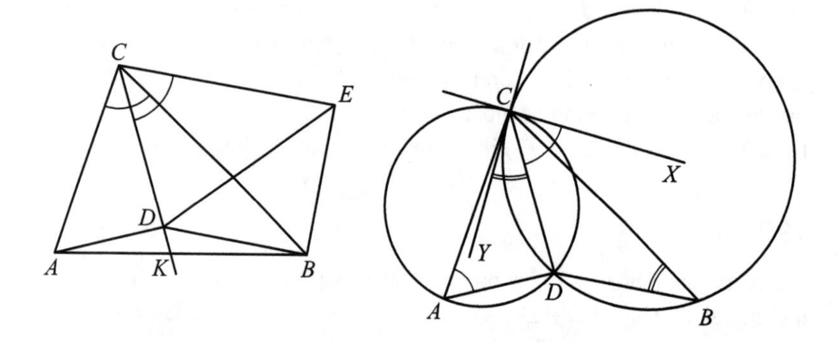

(b) Consider the circumcircles of triangles ACD and BCD. Let CX be the ray tangent to arc CD of the first circle and let CY be the ray tangent to arc CD of the second circle (X and Y are any points on those rays). Then $\angle DCX = \angle CAD$ and $\angle DCY = \angle CBD$ by the tangent-chord theorem.

Hence, according to (1), $\angle DCX + \angle DCY = 90°$. And since ray CD is situated inside the convex angle formed by rays CX and CY, we conclude that these last two rays are perpendicular, as claimed.

1993/3

Place a coordinate system on the plane so that the centers of the chessboard squares coincide with the integer lattice points. Color the squares in three colors; the square centered at (i, j) shall be given color 0, 1 or 2 according as the sum $i + j$ leaves remainder 0, 1 or 2 in division by 3.

Let k_0, k_1, k_2 be the numbers of *occupied* squares of the corresponding colors at a certain moment of the game. As an effect of any legal move, two of these numbers decrease by 1 and the remaining one increases by 1. Thus all the three numbers change their parities in each move.

If n is divisible by 3, then an $n \times n$ square contains equally many unit squares of each color, and so the initial values of k_0, k_1, k_2 are equal. At each moment of the game all the three numbers are simultaneously even or odd, so we will never achieve the state with one of them being equal to 1 and the other two 0; the game cannot end with just one piece on the board.

We are now going to prove that, conversely, if n is not divisible by 3, then the game can end with such a state. We devise an algorithm based on the following lemma.

Lemma. *Let $PQRS$ be a rectangle composed of $3\times m$ unit squares; $PQ = RS = 3$, $QR = SP = m$. Let* **x** *and* **z** *be the unit squares outwardly adjacent to sides PQ and RS of the rectangle, so that* **x** *has a vertex at P and* **z** *has a vertex at S. Suppose that the square* **x** *is unoccupied, the square* **z** *is occupied, and all the squares in the rectangle $PQRS$ are occupied (the state of the remaining squares in the board is irrelevant). Then it is possible to arrive at the situation in which the whole rectangle $PQRS$ is unoccupied, while the state of all the other squares in the board remains unaltered.*

Proof of the lemma. Label **u**, **v**, **w** the three squares of the rectangle, inwardly adjacent to side PQ, so that **u** has a vertex at P and **w** has a vertex at Q. Let **y** be the square adjacent to **u**, on the side opposite to **x**. Jump from **y** to **x**, killing the piece on square **u**. Then jump from **w** to **u**, killing the piece on square **v**. Finally, jump from **x** to **y**, killing the new piece on square **u**. The squares **u**, **v**, **w** have been cleared of pieces and the situation is similar to that from which we started, with a $3 \times (m-1)$ rectangle $P'Q'RS$ filled with pieces. Proceeding analogously, we are able to clear the whole rectangle $PQRS$; in the last step the role of square **y** is taken over by **z** (which is occupied by hypothesis). This proves the lemma.

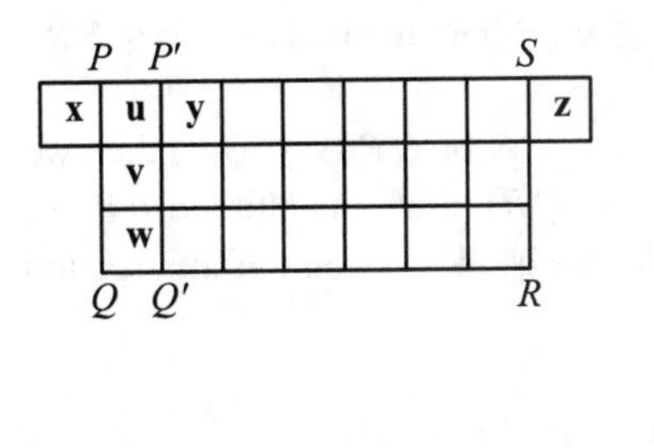

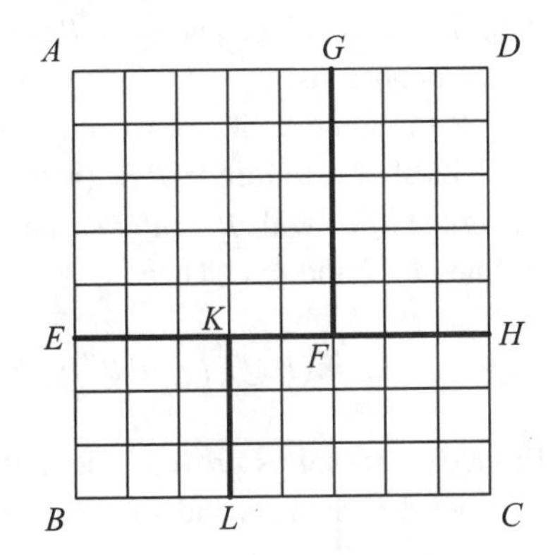

Now, assume that n is not divisible by 3 and let $ABCD$ be the $n \times n$ square occupied by pieces at the start. Let $AEFG$ be the $(n-3) \times (n-3)$ square with E on AB and G on AD, let line EF cut CD at H, and let $BLKE$ be the 3×3 square with L on BC and K on EH.

Apply the lemma three times, taking for $PQRS$, in succession, the following three rectangles: $DGFH$, $CHKL$ and $BLKE$; in each of these instances the conditions of the lemma are satisfied. The effect is that we get rid of all the pieces from the hexagon $BCDGFE$, and are left with the $(n-3) \times (n-3)$ square $AEFG$ filled with pieces and the rest of the chessboard free of pieces.

In an analogous fashion we reduce the occupied region to a square of side $n-6$, and so on. Eventually we obtain a 1×1 square or a 2×2 square filled with pieces (and the rest of the chessboard free). In the first case we are done; and in the second case it is no problem to kill three pieces, so that, again, the task is done.

Conclusion: the game can end with exactly one piece on the board if and only if n is not divisible by 3.

1993/4

We begin by showing that

$$\text{if} \quad T \in \triangle PQR, \quad \text{then} \quad m(PQT) \le m(PQR). \qquad (1)$$

(The notation $T \in \triangle PQR$ means: T is a point inside triangle PQR or on its boundary). To justify this observation, consider two cases. If PQ is the longest side of triangle PQR (hence also of triangle PQT), then the shortest altitudes in those triangles are the altitudes RR' and TT' dropped to PQ. The claimed inequality $m(PQT) \le m(PQR)$ reduces to $TT' \le RR'$, which is obvious.

If the longest side of triangle PQR is PR or QR (say, QR) then we draw its shortest altitude PP' (with P' on QR), draw the altitude PP'' of triangle PQT (with P'' on QT), and denote by W the point of intersection of lines PP' and QT. Then

$$m(PQT) \le PP'' \le PW \le PP' = m(PQR).$$

Property (1) is thus proved; notice that the reasoning does not depend on whether the triangles under consideration are acute-angled or not.

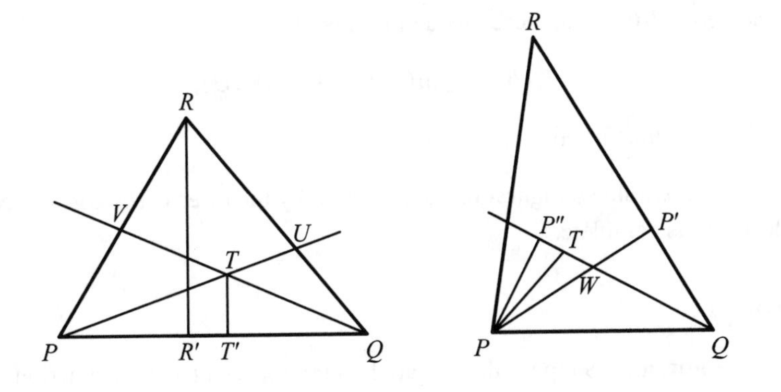

We pass to the solution proper; i.e., to the proof of

$$m(ABC) \leq m(ABX) + m(AXC) + m(XBC). \tag{2}$$

Case 1. X is a point inside or on the boundary of triangle ABC.

Denote this triangle by Δ_0 and the triangles XBC, AXC, ABX by Δ_1, Δ_2, Δ_3, respectively. Let a_i, h_i, S_i be the longest side, the shortest altitude, and the area of triangle Δ_i $(i = 0, 1, 2, 3)$. Clearly, a_0 is the maximum distance available between pairs of points in triangle Δ_0; in particular, $a_1, a_2, a_3 \leq a_0$. Since $a_i h_i = 2S_i$ and $S_0 = S_1 + S_2 + S_3$, we get

$$h_0 = \frac{a_1}{a_0}h_1 + \frac{a_2}{a_0}h_2 + \frac{a_3}{a_0}h_3 \leq h_1 + h_2 + h_3,$$

and this is nothing else than (2).

Case 2. One of the points A, B, C lies inside or on the boundary of the triangle spanned by the remaining two points and X.

Let e.g., $C \in \Delta ABX$. In view of property (1), $m(ABC) \leq m(ABX)$; claim (2) follows obviously.

Case 3. No one of the points A, B, C, X lies in the triangle spanned by the remaining three points.

Then these four points span a convex quadrilateral. Assume without loss of generality that $ABXC$ is a convex quadrilateral. Let X' be the point of intersection of its diagonals. According to observation (1),

$$m(ABX') \leq m(ABX) \quad \text{and} \quad m(ACX') \leq m(ACX);$$

and according to the conclusion of case 1 (with X' in place of X),

$$m(ABC) \leq m(ABX') + m(AX'C) + m(X'BC).$$

Since $m(X'BC) = 0$, these inequalities result in

$$m(ABC) \leq m(ABX) + m(AXC),$$

which of course implies (2).

All possible configurations are covered by the three cases considered; the proof is complete.

1993/5

First Solution. Suppose that we are looking for a *real-valued* function f, defined on N and satisfying the equation

$$f(f(n)) = f(n) + n \quad \text{for } n \in \mathrm{N}. \tag{1}$$

A possible start is to try a function of the form $f(n) = \alpha n$. Equation (1) then forces

$$\alpha^2 n = \alpha n + n \quad \text{for } n \in \mathrm{N}; \tag{2}$$

equivalently, $\alpha^2 - \alpha - 1 = 0$, a quadratic equation with the positive root $\alpha = (1 + \sqrt{5})/2$. In the sequel, α will denote this specific constant.

To make $(n \mapsto \alpha n)$ into an integer-valued function, we just round the value αn to the nearest integer:

$$f(n) = \lfloor \alpha n + \tfrac{1}{2} \rfloor \quad \text{for } n \in \mathrm{N}. \tag{3}$$

We are going to show that the function $f\colon \mathrm{N} \to \mathrm{N}$ thus obtained meets the requirements of the problem. Since α is irrational, we infer

$$|f(n) - \alpha n| < \tfrac{1}{2} \quad \text{for } n \in \mathrm{N}. \tag{4}$$

Replacing n by $f(n)$ we now get

$$|f(f(n)) - \alpha f(n)| < \tfrac{1}{2} \quad \text{for } n \in \mathrm{N}. \tag{5}$$

From equation (2) we obtain

$$\begin{aligned} f(f(n)) - f(n) - n &= f(f(n)) - f(n) - (\alpha^2 n - \alpha n) \\ &= (f(f(n)) - \alpha f(n)) + (\alpha - 1)\,(f(n) - \alpha n)\,. \end{aligned}$$

Hence, using the estimates (4) and (5),

$$|f(f(n)) - f(n) - n| < \tfrac{1}{2} + \tfrac{1}{2}(\alpha - 1) = \tfrac{1}{2}\alpha < 1;$$

and since $f(f(n)) - f(n) - n$ is a whole number, it must be equal to zero. This means that the function f, defined by (3), satisfies equation (1).

As for the other conditions from the problem statement, the equality $f(1) = 2$ follows directly from (3), and the inequality $f(n) < f(n+1)$ for $n \in \mathrm{N}$ follows from $\alpha > 1$, which guarantees that the numbers αn and $\alpha(n+1)$ are rounded to different integers.

Conclusion: functions with the properties under investigation do exist; the function (3) with the constant $\alpha = (1+\sqrt{5})/2$ is an example.

Second Solution. We will define two functions, $g\colon \mathrm{N} \setminus \{1\} \to \mathrm{N}$ and $f\colon \mathrm{N} \to \mathrm{N}$. To begin with, set $f(1) = 2$. Take an $n \in \mathrm{N}, n \geq 2$, and assume that the values $f(k)$ for $k = 1, \ldots, n-1$ have been already defined. Then let

$$g(n) = \max\{k\colon\ 1 \leq k < n, \quad f(k) \leq n\} \tag{6}$$

and

$$f(n) = n + g(n). \tag{7}$$

Since $f(1) = 2 \leq n$, the number 1 is an element of the set on the right side of (6); therefore $g(n) \geq 1$, and so, by (7), $f(n) > n$.

The functions g and f are now defined in their whole domains. From (6) it obviously follows that g is nondecreasing. Hence, by (7), f is a strictly increasing function—at least, in the set $\mathrm{N}\setminus\{1\}$; and since $f(2) = 2+g(2) = 2+1 > 2 = f(1)$, we see that f is increasing in the whole of N.

Fix $n \geq 2$ and denote $f(n)$ by m; then $m > n$. Consequently, the number n is an element of the set $\{k\colon 1 \leq k < m, f(k) \leq m\}$; in fact, it is the greatest element of this set, because the number $f(n+1)$ is greater than m $(= f(n))$. This means that $g(m) = n$, in agreement with the definition (6) (with m in place of n). And replacing n by m in (7), we get $f(m) = m + g(m)$; in other words, $f(f(n)) = f(n) + n$.

So we have equality (1); and since the function f is strictly increasing, we see that it satisfies all the conditions postulated.

Remark. If $f\colon \mathrm{N} \to \mathrm{N}$ is any function satisfying equation (1) with the initial condition $f(1) = 2$, then

$$f(2) = f(f(1)) = 1+2 = 3, \quad f(3) = f(f(2)) = 2+3 = 5,$$

and so on. In general, f is uniquely determined on the terms of the Fibonacci sequence:

$$f(u_j) = u_{j+1} \quad \text{for} \quad j = 0, 1, 2, \ldots,$$

where $u_0 = u_1 = 1$, $u_{j+1} = u_j + u_{j-1}$. The difficulty consists in how to fill in the gaps between the successive Fibonacci numbers; i.e., how to define the values $f(n)$ for $n \in (u_j, u_{j+1})$. As we have seen, there are several ways to overcome this difficulty, giving rise to *different* functions f. In particular, the two functions f, defined in the two solutions presented above, do not coincide. For instance, the value $f(9)$ equals 15 for the first one of those functions, and 14 for the second one.

1993/6

Let $x_j \in \{0, 1\}$ represent the state of lamp L_j after step S_j (0 for OFF and 1 for ON). Operation S_j affects the state of L_j, which in the previous round has been set to the value x_{j-n}. At the moment when S_j is being performed, lamp L_{j-1} is in state x_{j-1}. Consequently,

$$x_j \equiv x_{j-n} + x_{j-1} \pmod{2}. \tag{1}$$

This is true for all $j \geq n$, and also for $j = 0, \ldots, n-1$ if we agree that

$$x_{-n} = x_{-n+1} = x_{-n+2} = \cdots = x_{-2} = x_{-1} = 1, \tag{2}$$

which corresponds to the initial state: all lamps ON.

The state of the system at instant j can be represented by the vector $\mathbf{v}_j = [x_{j-n}, \ldots, x_{j-1}]$ (initial state: $\mathbf{v}_0 = [1, \ldots, 1]$). Since there are only 2^n feasible vectors, repetitions must occur in the sequence $\mathbf{v}_0, \mathbf{v}_1, \mathbf{v}_2, \ldots$. The operation that produces $\mathbf{v}_{j+1}$ from $\mathbf{v}_j$ is invertible. Therefore any recurrence equality $\mathbf{v}_{j+m} = \mathbf{v}_j$ forces $\mathbf{v}_m = \mathbf{v}_0$; the initial state recurs in at most 2^n steps, proving claim (a).

To prove (b) and (c), notice that in view of (1)

$$\begin{aligned} x_j &\equiv x_{j-n} + x_{j-1} \equiv (x_{j-2n} + x_{j-n-1}) + (x_{j-1-n} + x_{j-2}) \\ &\equiv x_{j-2n} + 2x_{j-n-1} + x_{j-2} \\ &\equiv x_{j-3n} + 3x_{j-2n-1} + 3x_{j-n-2} + x_{j-3}, \end{aligned}$$

and so on. After r applications of (1) we arrive at the equality

$$x_j \equiv \sum_{i=0}^{r} \binom{r}{i} x_{j-(r-i)n-i} \pmod{2},$$

holding for all j and r such that $j - (r-i)n - i \geq -n$. In particular, if r is of the form $r = 2^k$, then the binomial coefficients $\binom{r}{i}$ are even, except the two outer 1s (see the Remark), and we obtain

$$x_j \equiv x_{j-rn} + x_{j-r} \quad (\text{for } r = 2^k), \tag{3}$$

provided the subscripts do not go below $-n$; that means, for $j \geq (r-1)n$.

Now, if $n = 2^k$, choose $j \geq n^2 - n$ and set in (3) $r = n$, obtaining (in view of (1))

$$x_j \equiv x_{j-n^2} + x_{j-n} \equiv x_{j-n^2} + (x_j - x_{j-1}).$$

Hence $x_{j-n^2} = x_{j-1}$, showing that the sequence (x_j) is periodic with period $n^2 - 1$. Thus the string (2) of 1s reappears after exactly $n^2 - 1$ steps; claim (b) results.

And if $n = 2^k + 1$, choose $j \geq n^2 - 2n$ and set in (3) $r = n - 1$, to obtain (in view of (1))

$$\begin{aligned} x_j &\equiv x_{j-n^2+n} + x_{j-n+1} \equiv x_{j-n^2+n} + (x_{j+1} - x_j) \\ &\equiv x_{j-n^2+n} - x_{j+1} + x_j \end{aligned}$$

because $x \equiv -x \pmod 2$. Hence $x_{j-n^2+n} = x_{j+1}$, showing that the sequence (x_j) is periodic with period $n^2 - n + 1$ and proving claim (c).

Remark. In deriving formula (3) we have used the fact that for $r = 2^k$ the numbers $\binom{r}{1}, \binom{r}{2}, \ldots, \binom{r}{r-1}$ are even. We give a proof for completeness. Write each $j = 1, \ldots, r-1$ in the form $j = 2^{\alpha_j} m_j$ ($\alpha_j \geq 0, m_j \geq 1$ odd). Evidently, $\alpha_j < k$, as $j < r = 2^k$. Then $r - j = 2^{\alpha_j} q_j$ where $q_j = 2^{k-\alpha_j} - m_j$ is a positive integer. For $i = 1, \ldots, r-1$ we now get

$$\begin{aligned} \binom{r}{i} &= \frac{r \cdot (r-1) \cdots (r-i+1)}{i!} = r \left(\prod_{j=1}^{i-1} (r-j) \right) \left(\prod_{j=1}^{i} j \right)^{-1} \\ &= 2^k \left(2^{\alpha_1 + \cdots + \alpha_{i-1}} q_1 \cdots q_{i-1} \right) \left(2^{\alpha_1 + \cdots + \alpha_i} m_1 \cdots m_i \right)^{-1}. \end{aligned}$$

Hence $\binom{r}{i} m_1 \cdots m_i = 2^{k-\alpha_i} q_1 \cdots q_{i-1}$. The right side of the last equality is an even number, the product $m_1 \cdots m_i$ is odd, and therefore $\binom{r}{i}$ is even.

Thirty-fifth International Olympiad, 1994

1994/1

We may assume that the integers a_i are arranged increasingly: $a_1 < a_2 < \cdots < a_m$. Consider any pair of indices i, j such that $a_i + a_j \leq n$. By hypothesis, each one of the j sums $a_i+a_1, a_i+a_2, \ldots, a_i+a_{j-1}, a_i+a_j$ is equal to some a_k. All those a_ks are greater than a_i; their indices $k_1, k_2, \ldots, k_{j-1}, k_j$ form a j-element subset of $\{i+1, \ldots, m\}$; hence $m - i \geq j$. So we have shown that if $a_i + a_j \leq n$, then $i + j \leq m$. Consequently,

$$a_i + a_{m+1-i} \geq n+1 \quad \text{for } i = 1, 2, \ldots, m.$$

Summing these inequalities over $i = 1, 2, \ldots, m$ we eventually obtain

$$2(a_1 + a_2 + \cdots + a_m) \geq m(n+1);$$

and this is exactly what had to be proved.

1994/2

First Solution. We first assume that OQ is perpendicular to EF and show that then $QE = QF$.

The points B, E lie on one side of line OQ, and C, F lie on the other side. Consider the circles with diameters OE and OF. Since $OB \perp BE$ (and, by symmetry, $OC \perp CF$), point B lies on the first of these two circles, and C lies on the second. And since $OQ \perp EF$, point Q lies on both. Thus $\angle OBQ = \angle OEQ$ and $\angle OCQ = \angle OFQ$. Angles OBQ and OCQ are equal because OBC is an isosceles triangle. Hence $\angle OEQ = \angle OFQ$, showing that the right triangles OQE and OQF are congruent, and consequently $QE = QF$, as claimed.

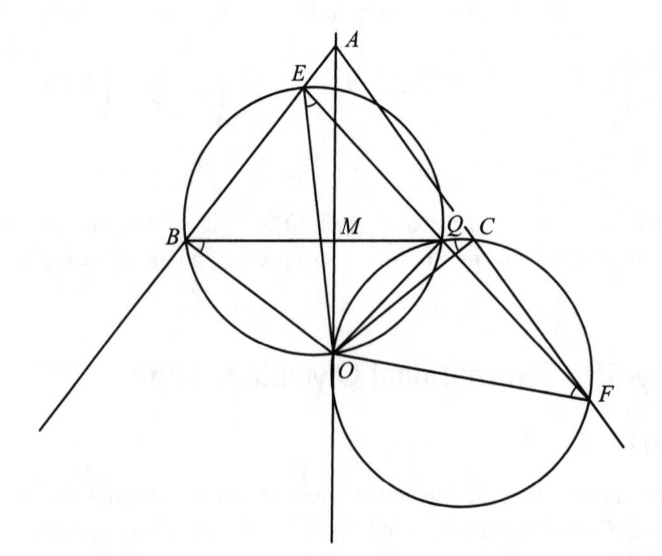

We are left with the proof of the converse implication. Assume that $QE = QF$ and draw through Q the line perpendicular to OQ, cutting AB and AC at the respective points E_0 and F_0. According to what has been already shown, $QE_0 = QF_0$. So Q is the common midpoint of segments EF and E_0F_0, and therefore the quadrilateral EE_0FF_0 is a parallelogram, unless it degenerates to a segment. And since the lines AB and AC are not parallel, EE_0FF_0 cannot be a nondegenerated parallelogram. Thus the points E_0 and F_0 coincide respectively with E and F, which means that the line EF is perpendicular to OQ. Done.

Second Solution. Let ω be the circle centered at O, tangent to AB and AC at B and C. Let K be the point symmetric to B, and L be the point symmetric to C across line OQ. The trapezoid $BKCL$ has ω as circumcircle; its sides BL and CK, extended, intersect at a point Q' on line OQ.

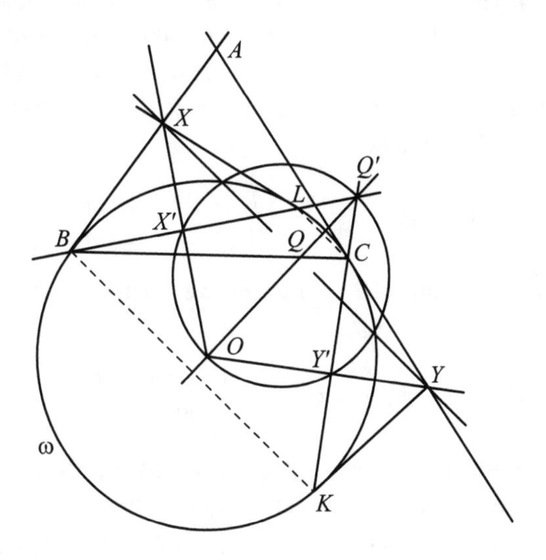

Let X and Y be points on AB and AC such that $OX \perp BL$ and $OY \perp CK$. Then OX cuts the segment BL at its midpoint X' and OY cuts the segment CK at its midpoint Y'. The isosceles triangles BXL and CYK are situated symmetrically with respect to line OQ; hence

$$QX = QY \quad \text{and} \quad OQ \perp XY. \tag{$*$}$$

Consider the inversion with respect to circle ω. Points Q and Q' (the intersections of lines BC, KL and BL, CK in the trapezoid $BKCL$)

are mutually inverse; also X, X' and Y, Y' are pairs of inverse points; these properties of inversion are well known (and not hard to prove). Since $OX'Q'$ and $OY'Q'$ are right triangles, the points Q', X' and Y' lie on a circle passing through O, the inversion center. Thus the inversion transforms this circle into a straight line passing through Q, X and Y.

The collinearity of these three points, together with relations $(*)$, show that either X and Y coincide respectively with E and F, and then the equality $QE = QF$ holds, or X and Y do not coincide with E and F, and then the equality $QE = QF$ fails to hold. The asserted equivalence results.

1994/3

Let us inspect how the value of f changes on passing from k to $k+1$. Denote by T the set of all integers whose binary representation has precisely three 1s. The value $f(k)$ counts the elements of T that are in the set $\{k+1, k+2, \ldots, 2k\}$, while $f(k+1)$ counts the elements of T in the set $\{k+2, \ldots, 2k, 2k+1, 2k+2\}$. Note that the numbers $k+1$ and $2k+2$ have equally many 1s in their representations; hence they simultaneously do or do not belong to T. Therefore

$$f(k+1) - f(k) = \begin{cases} 1 & \text{if } 2k+1 \in T, \\ 0 & \text{if } 2k+1 \notin T. \end{cases}$$

The binary representation of $2k+1$ arises from that of k by attaching a 1 at the end. So

$$f(k+1) - f(k) = \begin{cases} 1 & \text{if } k \text{ has exactly two 1s,} \\ 0 & \text{otherwise.} \end{cases} \tag{1}$$

The initial term of the sequence $f(1), f(2), f(3), \ldots$ is $f(1) = 0$. And since there are infinitely many integers k with exactly two 1s in their binary expansions, it follows from (1) that every positive integer m appears in this sequence; this is claim (a).

We also see from (1) that this sequence is piecewise constant. In order that an integer m be attained as a value $f(k)$ for exactly one k, it is necessary and sufficient that a jump should occur in passing from $k-1$ to k, as well as from k to $k+1$. According to (1), this is the case if and only if both $k-1$ and k have exactly two 1s.

The number $k-1$ cannot be even (in which case it would end with a 0, so that k would have one 1 more). Thus $k-1$ must be odd; its representation begins and ends with a 1, these two 1s being separated by a nonempty block of 0s (if this block were empty, $k-1$ would be $(11)_2$, and k would be $(100)_2$, with only one 1). Thus $k-1$ must be a number of the form 2^r+1 with $r \geq 2$; in other words:

$$k = 2^r + 2 \quad \text{where} \quad r \in \{2, 3, 4, \ldots\}. \tag{2}$$

And conversely, if k is any such number, then both $k-1$ and k have exactly two 1s. All that remains is to evaluate $m = f(k)$ for k of the form (2).

Note that

$$f(2^r) = \binom{r}{2} \quad \text{for } r = 2, 3, 4, \ldots; \tag{3}$$

this is so because for $k = 2^r$ the set $T \cap \{k+1, k+2, \ldots, 2k\}$ consists of those numbers whose binary representation has a leading 1 followed by r digits, with exactly two 1s among them; and of course, there are $\binom{r}{2}$ ways to choose their positions.

For each exponent $r \geq 2$ the number 2^r has one 1 and the number $2^r + 1$ has two 1s in its binary expansion. According to the rule (1),

$$f(2^r + 1) - f(2^r) = 0, \quad f(2^r + 2) - f(2^r + 1) = 1.$$

Adding these two equalities and using (3) we eventually conclude that for k as in (2),

$$m = f(2^r + 2) = f(2^r) + 1 = \binom{r}{2} + 1 \quad \text{where} \quad r \in \{2, 3, 4, \ldots\}.$$

These are all the values of m sought in part (b) of the problem.

1994/4

First Solution. Suppose that (m, n) is a "good" pair; this means that $(n^3+1)/(mn-1)$ is an integer. Dividing this integer by n we obtain quotient q and remainder r:

$$\frac{n^3+1}{mn-1} = qn + r; \quad q \geq 0, \quad 0 \leq r < n; \tag{1}$$

equivalently,

$$n^3 + 1 = (qn + r)(mn - 1) = qmn^2 + rmn - qn - r. \tag{2}$$

Hence $r \equiv -1 \pmod{n}$; i.e., $r = n - 1$, so that (2) becomes

$$n^2 = qmn + m(n-1) - q - 1. \tag{3}$$

Note that if (m, n) is a "good" pair, then the number

$$m^3 \cdot \frac{n^3+1}{mn-1} = \frac{m^3n^3-1}{mn-1} + \frac{m^3+1}{mn-1}$$

is an integer; hence $(m^3+1)/(mn-1)$ is an integer and (n, m) is also a "good" pair. So we may restrict attention to pairs (m, n) with $m \geq n$.

If $n = 1$, we get that $2/(m-1)$ is an integer, implying that m is either 2 or 3; so we have the first two "good" pairs (2, 1) and (3, 1).

For the sequel assume $m \geq n \geq 2$. Going back to formula (3) we obtain $n^2 \geq qn^2 + n(n-1) - q - 1$; after recasting, $q(n-1) \leq 1$. Thus $q = 0$ or $q = 1$. In the latter case we must have $n = 2$, with equality in the previous estimate; i.e., $m = n = 2$. So we have another "good" pair (2, 2).

In the remaining case ($q = 0$, $n \geq 2$), equation (3) is transformed into

$$m = n + 1 + \frac{2}{n-1},$$

implying that n is either 2 or 3; in each case $m = 5$. This yields two more "good" pairs (5, 2) and (5, 3).

Taking into account the symmetry between m and n we obtain, in all, nine pairs (m, n) satisfying the imposed condition:

$$(2, 1), (1, 2), (3, 1), (1, 3), (2, 2), (5, 2), (2, 5), (5, 3), (3, 5).$$

Second Solution. We begin as in the first solution and arrive at equation (3). Denoting $q + 1$ by k we rewrite (3) as $n^2 = kmn - k - m$. So n is a root of the quadratic equation

$$x^2 - kmx + (k + m) = 0.$$

Let ℓ be the other real root (we do not exclude the possibility that $\ell = n$). The sum and the product of the roots are expressed by the familiar formulas

$$\ell + n = km, \quad \ell n = k + m. \tag{4}$$

The numbers k, m, n are positive integers; thus ℓ is also a positive integer. Summing the two equations of (4) we obtain $km + \ell n = \ell + n + k + m$, which we rewrite in the form

$$u + v = 2 \quad \text{where} \quad u = (k-1)(m-1), \quad v = (\ell - 1)(n - 1).$$

Note that u and v are nonnegative integers; so we have either $u = v = 1$, implying

$$k = m = \ell = n = 2, \tag{5}$$

or one of the numbers u, v is 0 and the other is 2.

Assume $u = 2$; then one of the factors in $(k-1)(m-1)$ must be 1 and the other 2; i.e., one of the numbers k, m is 2 and the other is 3. The system (4) takes the form $\ell + n = 6$, $\ell n = 5$; one of the numbers ℓ, n is 1 and the other is 5. So we get another four solutions (k, m, ℓ, n) of the equation system (4):

$$(2, 3, 1, 5), \quad (2, 3, 5, 1), \quad (3, 2, 1, 5), \quad (3, 2, 5, 1). \tag{6}$$

In the remaining case, i.e., when $v = 2$, the situation is quite analogous, in view of the symmetry between the pairs $\{k, m\}$ and $\{\ell, n\}$ in the equation system (4). We obtain the following solutions (k, m, ℓ, n):

$$(1, 5, 2, 3), \quad (1, 5, 3, 2), \quad (5, 1, 2, 3), \quad (5, 1, 3, 2). \tag{7}$$

From (5), (6), (7) we see (rejecting the auxiliary unknowns k and ℓ) that (m, n) must be one of the nine pairs found in the first solution, and we check directly that each one of those pairs indeed satisfies the required condition.

1994/5

Suppose that $f\colon S \to S$ is a function satisfying the given conditions. Note that for every $x, y \in S$ the number $x + f(y) + xf(y)$ (appearing as the argument of f in condition (a)) can be written as $(x+1)(f(y)+1) - 1$, and hence belongs to S; so the problem is correctly posed. Setting $x = y$ in condition (a) we obtain the equation

$$f(x + f(x) + xf(x)) = x + f(x) + xf(x) \quad \text{for } x \in S. \tag{1}$$

Fix an $x \in S$ and denote the number $x + f(x) + xf(x)$ by s. So $f(s) = s$; in other words, s is a fixed point of the function f. Applying equation (1) with s in place of x we get that also the number $t = s + f(s) + sf(s)$ is a fixed point of f.

If $s \neq 0$, then the number $t = s + f(s) + sf(s) = 2s + s^2 = s(s+2)$ is different from s and has the same sign as s. Both s and t satisfy the equation $f(x)/x = 1$. This contradicts condition (b) which says that the function $f(x)/x$ is strictly increasing on each of the intervals $(-1, 0)$ and $(0, \infty)$. Therefore s must be zero: $x + f(x) + xf(x) = 0$. Solving this for $f(x)$ we obtain

$$f(x) = -\frac{x}{x+1}. \tag{2}$$

Since $x \in S$ was chosen arbitrarily, the last formula represents the only function that can be a solution of the problem. It is a matter of simple verification that this function indeed satisfies both conditions (a) and (b).

Remark. The given conditions are wasteful, as can be seen from the solution above. To determine f uniquely, weaker conditions are sufficient; namely, the functional equation (1) (with only one independent variable, instead of the original equation involving two variables) plus the information that each one of the intervals $(-1, 0)$ and $(0, \infty)$ can contain at most one fixed point of the function f.

1994/6

First Solution. This is an exercise about square-free numbers. An integer x is called *square-free* if it is not divisible by the square of any integer greater than 1.

Let $p_1, p_2, p_3, \dots$ be the increasing sequence of all prime numbers ($p_1 = 2$, $p_2 = 3$, $p_3 = 5$ etc.). An integer $x \geq 2$ is square-free if and only if it can be written in the form

$$x = p_{i_1} p_{i_2} \cdots p_{i_k} \quad \text{where} \quad i_1 < i_2 < \cdots < i_k. \tag{1}$$

In the construction of the set A, only square-free numbers are of any significance; no other number can appear in the role of m or n from the problem statement. Integers that are not square-free can be included into A or not, as we please. Let us agree, just for definiteness, that we do not include them in A.

The "trouble" consists in the fact that there is a lot of freedom in the construction of A. We will show two methods of construction; but there are a great many others; we invite the reader to devise some.

We propose the following definition: a square-free integer $x \geq 2$ of the form (1) belongs to A if and only if $i_1 = k$.

In other words: if p_k is the smallest prime divisor of x, then x, in order to belong to A, should have exactly k prime divisors in all.

Let S be any infinite set of primes and let p_k be the least odd prime in S; so $k \geq 2$. Let $i_1 = k$. Choose from S arbitrarily $2k - 1$ distinct primes $p_{i_2}, p_{i_3}, \ldots, p_{i_{2k}}$ greater than $p_{i_1} (= p_k)$ and consider the two numbers $m = p_{i_1} p_{i_2} \cdots p_{i_k}$ and $n = p_{i_{k+1}} p_{i_{k+2}} \cdots p_{i_{2k}}$. Then m satisfies the condition defining the set A, whereas n does not. So m and n are integers just as needed.

Second Solution. The beginning is as in the previous solution (till the "invitation for the reader"); we only propose a different definition of A: let us agree that a square-free integer $x \geq 2$ of the form (1) belongs to A if and only if

$$i_1 \equiv i_2 \equiv \cdots \equiv i_k \pmod{k}. \tag{2}$$

Let S be any infinite set of primes and let p_j, p_k ($j < k$) be the two smallest numbers in S; hence $k \geq 2$.

Choose arbitrarily $k(k-1) + 1$ elements $p_{s_1}, p_{s_2}, \ldots, p_{s_{k(k-1)+1}} \in S$. Each label s_r gives in division by k one of k possible remainders. By the pigeonhole principle, at least one of those remainders occurs at least k times. So we may choose k labels $i_1, \ldots, i_k$ to make condition (2) fulfilled; the number $m = p_{i_1} p_{i_2} \cdots p_{i_k}$ belongs to A.

Note that $j \not\equiv k \pmod{k}$. Thus if we take n to be the product of p_j, p_k and $k - 2$ other primes from S, we obtain an integer that does not fulfill condition (2), hence does not belong to A. This means that the set A defined by condition (2) has the required property.

Thirty-sixth International Olympiad, 1995

1995/1

First Solution. We may assume that P and X are on the same side of line AD. Denote the circle with diameter AC by ω_1 and the circle with diameter BD by ω_2. Let line DN intersect XY at V and let AM intersect XY at U. Our task is to show that U coincides with V.

Obviously, BND and BXD are right angles (subtended by the diameter of ω_2). So the sides of triangle BZP are perpendicular to the respective sides of triangle VZD; the triangles are similar and therefore $ZV/ZD = ZB/ZP$. Since XZ is the altitude in the right triangle BXD, we obtain $XZ^2 = ZB \cdot ZD = ZV \cdot ZP$.

Analogously, $XZ^2 = ZA \cdot ZC = ZU \cdot ZP$. It follows from these equalities that the points U and V coincide, and consequently the lines AM, DN, XY are concurrent (note that the solution is case-independent).

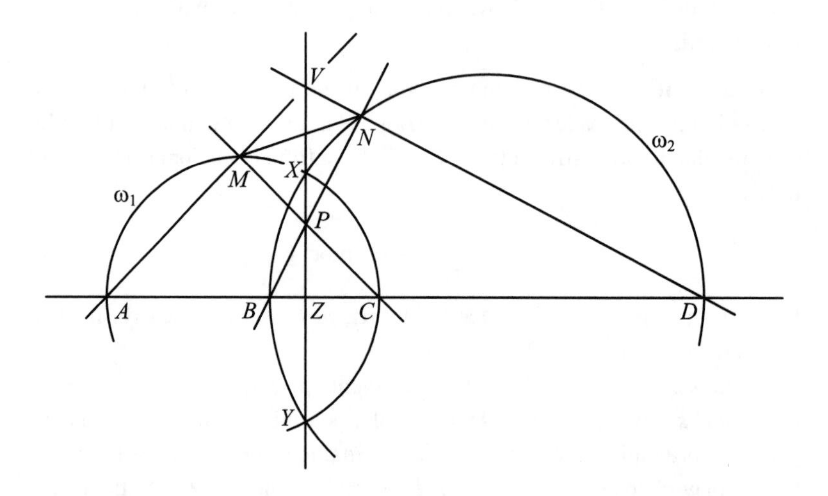

Second Solution. Again we assume that P and X lie on the same side of line AD. Moreover, assume that P lies between Z and X (as in the diagram; if X lies between Z and P, the formulas below have to be slightly modified; we leave the details to the reader).

Note that P has equal powers with respect to circles ω_1 and ω_2; i.e., we have the equalities

$$PC \cdot PM = PX \cdot PY = PB \cdot PN,$$

showing that the points B, C, M, N are concyclic; so $\angle MNB = \angle MCB$. Since AMC is a right triangle, $\angle MCB = \angle MCA = 90^\circ - \angle MAD$, and we get $\angle MNB = 90^\circ - \angle MAD$. Similarly, BND is a right triangle, so that

$$\angle MND = \angle MNB + \angle BND = \angle MNB + 90^\circ.$$

These equalities result in $\angle MND = 180^\circ - \angle MAD$. Consequently, the points A, M, N, D lie on a circle; denote this circle by ω_3.

Now, line AM is the radical axis of circles ω_1 and ω_3, line DN is the axis of circles ω_2 and ω_3, and line XY is the axis of circles ω_1 and ω_2. The axes of three circles, taken in pairs, are concurrent lines (see *Glossary*); hence the result.

1995/2

First Solution. Basically, all approaches to this inequality start from the substitution $a = 1/x$, $b = 1/y$, $c = 1/z$. The new variables x, y, z also take positive values and satisfy the condition $xyz = 1$. By the AM-GM inequality, $x + y + z \geq 3$. Since

$$\frac{1}{a^3(b+c)} = \frac{x^3}{(1/y) + (1/z)} = \frac{x^3yz}{y+z} = \frac{x^2}{y+z},$$

the result will be proved if we show that

$$\frac{x^2}{y+z} + \frac{y^2}{z+x} + \frac{z^2}{x+y} \geq \frac{x+y+z}{2}. \tag{1}$$

Note that (1) is a homogeneous inequality; we will prove that it holds for arbitrary positive real numbers x, y and z.

In the Cauchy–Schwarz inequality

$$(kp + lq + mr)^2 \leq (k^2 + l^2 + m^2)(p^2 + q^2 + r^2)$$

(see *Glossary*) set

$$k = \sqrt{y+z}, \quad l = \sqrt{z+x}, \quad m = \sqrt{x+y},$$
$$p = \frac{x}{k}, \quad q = \frac{y}{l}, \quad r = \frac{z}{m},$$

to obtain

$$(x+y+z)^2 \leq (2x + 2y + 2z)\left(\frac{x^2}{y+z} + \frac{y^2}{z+x} + \frac{z^2}{x+y}\right);$$

the claimed inequality (1) follows immediately.

Second Solution. As in the first solution, we begin by reducing the assertion to inequality (1). For any positive numbers α, β, γ,

$$\begin{aligned}(\alpha+\beta+\gamma)\left(\frac{1}{\alpha} + \frac{1}{\beta} + \frac{1}{\gamma}\right) &= 3 + \left(\frac{\alpha}{\beta} + \frac{\beta}{\alpha}\right) + \left(\frac{\beta}{\gamma} + \frac{\gamma}{\beta}\right) + \left(\frac{\gamma}{\alpha} + \frac{\alpha}{\gamma}\right) \\ &\geq 9.\end{aligned}$$

Setting $\alpha = y + z$, $\beta = z + x$, $\gamma = x + y$ (so $\alpha + \beta + \gamma = 2(x + y + z)$), we rewrite the last inequality as

$$(x+y+z)\left(\frac{1}{y+z} + \frac{1}{z+x} + \frac{1}{x+y}\right) \geq \frac{9}{2}. \tag{2}$$

Since

$$\frac{x^2}{y+z} = \frac{(x+y+z)^2}{y+z} - (2x+y+z),$$

we obtain the asserted inequality (1):

$$\begin{aligned}\frac{x^2}{y+z} + \frac{y^2}{z+x} + \frac{z^2}{x+y} &= (x+y+z)^2\left(\frac{1}{y+z} + \frac{1}{z+x} + \frac{1}{x+y}\right) \\ &\quad - 4(x+y+z) \\ &\geq (x+y+z)\left(\frac{9}{2} - 4\right) \\ &= \frac{x+y+z}{2}.\end{aligned}$$

Third Solution. Once more we reduce the result to inequality (1), as in the first solution. Assume without loss of generality $x \geq y \geq z$. Further reasoning will be based on formula (2) from the second solution and on the well-known inequality

$$xu + yv + zw \geq \frac{(x+y+z)(u+v+w)}{3}, \tag{3}$$

holding for any real numbers $x \geq y \geq z$ and $u \geq v \geq w$ (see the Remark). Setting

$$u = \frac{x}{y+z}, \quad v = \frac{y}{z+x}, \quad w = \frac{z}{x+y}$$

and using (2) we have

$$\begin{aligned}u+v+w &= \frac{x}{y+z} + \frac{y}{z+x} + \frac{z}{x+y} \\ &= \left(\frac{x+y+z}{y+z} - 1\right) + \left(\frac{x+y+z}{z+x} - 1\right) + \left(\frac{x+y+z}{x+y} - 1\right) \\ &= (x+y+z)\left(\frac{1}{y+z} + \frac{1}{z+x} + \frac{1}{x+y}\right) - 3 \geq \frac{9}{2} - 3 = \frac{3}{2}.\end{aligned}$$

The numbers u, v, w satisfy the condition $u \geq v \geq w$, which allows us to apply inequality (3):

$$\frac{x^2}{y+z} + \frac{y^2}{z+x} + \frac{z^2}{x+y} \geq \frac{(x+y+z)(u+v+w)}{3} \geq \frac{x+y+z}{2};$$

and this is (1).

Remark. Relation (3) is an instant of Chebyshev's inequality

$$x_1u_1 + x_2u_2 + \cdots + x_nu_n \geq \frac{(x_1 + x_2 + \cdots + x_n)(u_1 + u_2 + \cdots + u_n)}{n},$$

holding for $x_1 \geq x_2 \geq \cdots \geq x_n$, $u_1 \geq u_2 \geq \cdots \geq u_n$. For $n = 3$ we get inequality (3), whose proof we now outline for completeness: the difference between the left side of (3) and the right side of (3), multiplied by 3, can be transformed into the form

$$(x - y)(u - v) + (y - z)(v - w) + (z - x)(w - u),$$

which is of course nonnegative.

Fourth Solution. Again, we approach the problem in its reduced form (1); see the first solution. For a fixed value of the sum $s = x + y + z$ consider the function

$$f(t) = \frac{t^2}{s - t} \quad \text{for } t \in (0, s).$$

Its derivative

$$f'(t) = \frac{2t(s - t) + t^2}{(s - t)^2} = 2\left(\frac{s}{s - t} - 1\right) + \left(\frac{s}{s - t} - 1\right)^2$$

is an increasing function of variable t in the interval $(0, s)$. Thus f is a convex function in that interval, and hence it satisfies Jensen's inequality

$$f(x) + f(y) + f(z) \geq 3f\left(\frac{x + y + z}{3}\right) = 3f\left(\frac{s}{3}\right).$$

The last expression has value $s/2$, so we have obtained inequality (1).

1995/3

For $n = 4$, the vertices of a unit square $A_1A_2A_3A_4$ and the four equal numbers $r_1 = r_2 = r_3 = r_4 = 1/6$ meet the demands.

Take an integer $n \geq 5$ and suppose there exist points $A_1, A_2, \ldots, A_n$ and numbers $r_1, r_2, \ldots, r_n$ satisfying conditions (i) and (ii). The following observation will be the clue to further reasoning:

$$\text{if } A_iA_jA_kA_l \text{ is a convex quadrilateral, then } r_i + r_k = r_j + r_l. \quad (1)$$

Indeed: $\text{area}(A_iA_jA_k) + \text{area}(A_iA_kA_l) = \text{area}(A_iA_jA_l) + \text{area}(A_jA_kA_l)$. So $(r_i + r_j + r_k) + (r_i + r_k + r_l) = (r_i + r_j + r_l) + (r_j + r_k + r_l)$, according to condition (ii), and hence $r_i + r_k = r_j + r_l$; claim (1) is justified.

Consider the points A_1, A_2, A_3, A_4, A_5 and the smallest convex polygon containing them; it can be a pentagon, a quadrilateral or a triangle. We will consider these three cases separately.

Suppose $A_1A_2A_3A_4A_5$ is a convex pentagon. Applying observation (1) to quadrilaterals $A_1A_2A_3A_4$ and $A_1A_2A_3A_5$ we obtain the equalities

$$r_1 + r_3 = r_2 + r_4 \quad \text{and} \quad r_1 + r_3 = r_2 + r_5; \tag{2}$$

hence $r_4 = r_5$ and

$$\text{area}(A_1A_2A_4) = r_1 + r_2 + r_4 = r_1 + r_2 + r_5 = \text{area}(A_1A_2A_5),$$
$$\text{area}(A_2A_3A_4) = r_2 + r_3 + r_4 = r_2 + r_3 + r_5 = \text{area}(A_2A_3A_5).$$

The first equality shows that $A_4A_5 \parallel A_1A_2$; and the second one implies $A_4A_5 \parallel A_2A_3$. Points A_1, A_2, A_3 would be collinear, contrary to condition (i).

Now assume that four of the points A_1, A_2, A_3, A_4, A_5 span a quadrilateral containing the fifth point in its interior. Let e.g., $A_1A_2A_3A_4$ be a convex quadrilateral; without loss of generality we may assume that A_5 lies inside triangle $A_1A_3A_4$. We apply property (1) to quadrilaterals $A_1A_2A_3A_4$ and $A_1A_2A_3A_5$, obtaining as before the equalities (2). Thus $r_4 = r_5$. But this is impossible, in view of

$$r_1 + r_3 + r_4 = \text{area}(A_1A_3A_4) > \text{area}(A_1A_3A_5) = r_1 + r_3 + r_5. \tag{3}$$

It remains to consider the case in which three points, say, A_1, A_2, A_3, are the vertices of a triangle with points A_4 and A_5 inside it. Again, there is no loss of generality in assuming that A_5 lies inside triangle $A_1A_3A_4$. Inequality (3) is then true. However, for each one of the labels $k = 4$ and $k = 5$ we have

$$\begin{aligned}\text{area}(A_1A_2A_3) &= \text{area}(A_1A_2A_k) + \text{area}(A_1A_3A_k) + \text{area}(A_2A_3A_k)\\ &= (r_1 + r_2 + r_k) + (r_1 + r_3 + r_k) + (r_2 + r_3 + r_k),\end{aligned}$$

implying that r_k is the same for $k = 4$ and $k = 5$, in opposition to inequality (3).

In each case we have arrived at a contradiction. Thus $n = 4$ is the only integer for which conditions (i) and (ii) can be realized.

1995/4

Suppose $x_0, x_1, \ldots, x_{1995}$ is a sequence satisfying the given conditions. The equation from condition (ii) can be recast and factored into

$$(x_{i-1}x_i - 1)(x_{i-1} - 2x_i) = 0.$$

Thus, for each $i = 1, \dots, 1995$,

$$\text{either} \quad x_i = \frac{1}{x_{i-1}} \quad \text{or} \quad x_i = \frac{x_{i-1}}{2}. \qquad (*)$$

We claim that at least one x_i is equal to 1. Suppose not. Then all the x_is lie in the union of disjoint intervals

$$\dots \cup [\tfrac{1}{8}, \tfrac{1}{4}) \cup [\tfrac{1}{4}, \tfrac{1}{2}) \cup [\tfrac{1}{2}, 1) \cup (1, 2] \cup (2, 4] \cup (4, 8] \cup \dots.$$

Color these intervals alternately: ..., red, blue, red, blue, ... It is seen from $(*)$ that, in each pair of consecutive x_is, one term belongs to a red interval and the other to a blue one. Thus x_0 and x_{1995} are in intervals of different colors, in contradiction to condition (i). Thus, indeed, there exists k such that $x_k = 1$.

It follows from $(*)$ that if a term x_i belongs to an interval of the form $[a^{-1}, a]$ (for a certain positive number a), then x_{i-1} and x_{i+1} are in the interval $[(2a)^{-1}, 2a]$. The term $x_k = 1$ is the single element of the (degenerated) interval $[1, 1]$. Thus $x_{k\pm1} \in [\frac{1}{2}, 2]$, $x_{k\pm2} \in [\frac{1}{4}, 4]$, and by induction

$$x_{k-j}, x_{k+j} \in [2^{-j}, 2^j]$$

for all j such that x_{k-j} and x_{k+j} are defined; i.e., for j satisfying the inequalities $j \le k$ and $j \le 1995 - k$. Taking for j these two extreme values we get

$$x_0 \le 2^k \quad \text{and} \quad x_0 = x_{1995} \le 2^{1995-k}.$$

Hence $x_0 \le 2^m$ where $m = \min\{k, 1995 - k\} \le 997$.

The equality $x_0 = 2^{997}$ can be attained if we set

$$x_i = 2^{997-i} \quad \text{for } i = 0, 1, \dots, 1994 \quad \text{and} \quad x_{1995} = x_0;$$

equations $(*)$ are fulfilled. Thus 2^{997} is the maximum value of x_0.

1995/5

First Solution. By the conditions of the problem, triangles BCD and EFA are equilateral. Therefore $AB = DB$, $DE = AE$ and consequently triangles ABE and DBE are symmetric with respect to BE.

Let K be the point symmetric to C and L be the point symmetric to F across line BE. Clearly, $KL = CF$. Triangles BKA and ELD (symmetric to BCD and EFA) are equilateral.

On segments BG and EH erect equilateral triangles BGM and EHN with M lying on the same side of line BG as point K, and N on the same side of EH as L. Angles ABG and KBM are equal; this, in view of $AB = KB$ and $BG = BM$, implies that the triangles ABG and KBM are congruent. Thus $AG = KM$. Similarly, $DH = LN$. Thus, finally,

$$AG + GB + GH + HE + DH = KM + MG + GH + HN + NL \\ \geq KL = CF.$$

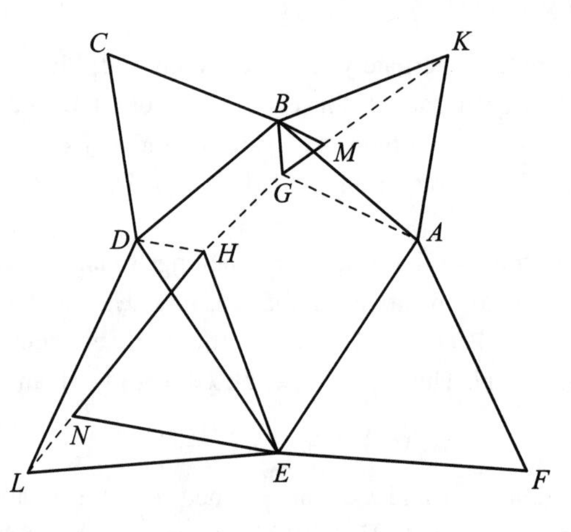

Second Solution. As in the first solution we find out that ABE and DBE are symmetric triangles and define K, L as points symmetric respectively to C, F across line BE; thus $KL = CF$. In each one of the convex quadrilaterals $AKBG$ and $DLEH$, Ptolemy's inequality holds:

$$AB \cdot KG \leq BK \cdot AG + KA \cdot GB, \quad DE \cdot HL \leq EL \cdot DH + LD \cdot HE$$

(see *Glossary*). In the first inequality, AB, BK, KA are equal segments, and in the second, DE, EL, LD are equal segments. Cancelling their lengths we get

$$KG \leq AG + GB \quad \text{and} \quad HL \leq DH + HE.$$

Consequently,

$$AG + GB + GH + DH + HE \geq KG + GH + HL \geq KL = CF.$$

Remark. The condition about the size of angles AGB and DHE, given in the problem statement, has not been used in any of these solutions; it is redundant.

1995/6

First Solution. The set $\{1, 2, \ldots, 2p\}$ is the union of two disjoint sets, $X = \{1, 2, \ldots, p\}$ and $Y = \{p+1, p+2, \ldots, p+p\}$. The sum of the elements of X equals $p(p+1)/2$ and is divisible by p (because p is odd); also the sum of the elements of Y is divisible by p.

Choose any p-element set $A \subset \{1, 2, \ldots, 2p\}$. Denote by k_A the number of elements of A in $\{1, \ldots, p\}$ and by r_A the remainder left by the sum of *all* elements of A in division by p. Clearly, $A = Y$ is the only set with $k_A = 0$ and $A = X$ is the only set with $k_A = p$.

Now let A be a p-element set other than X or Y; thus $0 < k_A < p$. There is exactly one integer $\ell_A \in \{1, \ldots, p-1\}$ such that

$$k_A \cdot \ell_A \equiv 1 \pmod{p},$$

because the multiples $k_A \cdot j$ for $j = 1, \ldots, p-1$ have different nonzero remainders in division by p.

With any such set A we associate the p-element set A^* defined as follows: the portion of A in X is shifted by ℓ_A cyclically within X, and the portion of A in Y is left unchanged; in symbols, $A^* = \{x^* : x \in A\}$, where

$$x^* = \begin{cases} x + \ell_A & \text{if } x + \ell_A \leq p, \\ x + \ell_A - p & \text{if } x \leq p < x + \ell_A, \\ x & \text{if } x > p. \end{cases}$$

To distinct sets A there correspond distinct sets A^*. On passing from A to A^*, the increment (modulo p) of the sum of numbers in the set equals $k_A \cdot \ell_A$. So

$$r_{A^*} \equiv r_A + k_A \cdot \ell_A \equiv r_A + 1 \pmod{p}.$$

Thus, for each $t \in \{0, 1, \ldots, p-1\}$, the operation $A \mapsto A^*$ establishes a bijection between the class of all p-element sets $A \subset \{1, 2, \ldots, 2p\}$, different from X and Y, such that $r_A = t$, and the analogous class with $r_A = t+1$ (which has to be understood as 0 when $r_A = p-1$). It follows that all these p classes are equipotent.

The union of these p classes (as $t \in \{0, 1, \ldots, p-1\}$) exhausts the family of all p-element subsets of $\{1, 2, \ldots, 2p\}$ other than X and Y. We have $n = \binom{2p}{p} - 2$ sets in this family, and so there are n/p sets in each one of those p classes. In particular, n/p is the number of all p-element sets, other than X or Y, with the sum of elements divisible by p. Including these two "trivial" sets back into consideration, we obtain the final outcome

$$\frac{1}{p}\left(\binom{2p}{p}-2\right)+2.$$

Second Solution. We proceed by the technique of polynomial algebra in the complex domain. Let $\epsilon = \cos(2\pi/p) + i\sin(2\pi/p)$. Then the numbers $1, \epsilon, \epsilon^2, \epsilon^3, \ldots, \epsilon^{p-1}$ are distinct pth roots of unity. Let ω be any one of these roots, other than $\epsilon^0 = 1$; i.e., let

$$\omega = \epsilon^k \quad \text{for a certain } k \in \{1, 2, \ldots, p-1\}. \tag{1}$$

Then the numbers

$$\omega^0 = 1, \quad \omega^1 = \epsilon^k, \quad \omega^2 = \epsilon^{2k}, \quad \ldots, \quad \omega^{p-1} = \epsilon^{(p-1)k} \tag{2}$$

are also pth roots of unity; they are distinct because p is a prime and hence the multiples jk for $j = 1, \ldots, p-1$ give different remainders in division by p; compare the analogous passage in the first solution. In other words, the sequence (2) is a permutation of the sequence $1, \epsilon, \epsilon^2, \ldots, \epsilon^{p-1}$.

Consider the polynomial

$$f(z) = z^p - 1 = \prod_{j=0}^{p-1}(z-\epsilon^j) = \prod_{j=0}^{p-1}(z-\omega^j), \tag{3}$$

together with its square $F(z) = f(z)^2$, which can be written as

$$F(z) = \prod_{j=1}^{p}(z-\omega^j)\cdot\prod_{j=p+1}^{2p}(z-\omega^j) = \prod_{j=1}^{2p}(z-\omega^j); \tag{4}$$

or, in the expanded form,

$$F(z) = a_0 + a_1 z + a_2 z^2 + \cdots + a_p z^p + \cdots + a_{2p-1}z^{2p-1} + z^{2p}.$$

Let us examine the coefficient a_p. The terms with z^p are obtained by arbitrarily choosing p pairs of parentheses in the factorization (4) and picking "the z term" from each of them; from any one of the remaining p pairs of parentheses we pick the corresponding term $-\omega^j$. The result is

$$a_p = \sum_{1\le j_1<j_2<\cdots<j_p\le 2p}\left(-\omega^{j_1}\right)\left(-\omega^{j_2}\right)\cdots\left(-\omega^{j_p}\right) = -\sum_{r=0}^{p-1} c_r\omega^r \tag{5}$$

where c_r denotes the number of those p-tuples $j_1 < j_2 < \cdots < j_p$ in which $j_1 + j_2 + \cdots + j_p \equiv r \pmod{p}$. Note that the sum of all the c_rs is the number of all p-element subsets of the set $\{1, 2, \dots, 2p\}$,

$$c_0 + c_1 + \cdots + c_{p-1} = \binom{2p}{p}, \tag{6}$$

while c_0 is the number of p-element subsets with the sum of elements divisible by p; i.e., the number we are about to evaluate.

Since $F(z) = f(z)^2 = z^{2p} - 2z^p + 1$, the coefficient a_p is equal to -2; equality (5) takes the form $c_0 + c_1\omega + c_2\omega^2 + \cdots + c_{p-1}\omega^{p-1} = 2$. In other words, ω is a root of the polynomial

$$g(z) = (c_0 - 2) + c_1 z + c_2 z^2 + \cdots + c_{p-1} z^{p-1}. \tag{7}$$

The definition of the numbers $c_0, c_1, \dots, c_{p-1}$—hence also of the polynomial $g(z)$—was purely combinatorial and independent of the choice of a particular number ω of the form (1). Therefore each one of those numbers $(\epsilon, \epsilon^2, \dots, \epsilon^{p-1})$ is a root of $g(z)$. Since $g(z)$ has degree $p - 1$, we get in view of (3)

$$\begin{aligned} g(z) &= c_{p-1} \prod_{j=1}^{p-1} (z - \epsilon^j) = c_{p-1} \cdot \frac{z^p - 1}{z - 1} \\ &= c_{p-1}(1 + z + \cdots + z^{p-1}). \end{aligned} \tag{8}$$

It follows that all the coefficients of the polynomial (7) are equal:

$$c_0 - 2 = c_1 = c_2 = \cdots = c_{p-1}.$$

According to (6), their sum is $\binom{2p}{p} - 2$. Thus, eventually,

$$c_0 = 2 + \frac{1}{p}\left(\binom{2p}{p} - 2\right).$$

Remark. With a little more knowledge of algebra, the write-up of the second solution can be made more concise. There is, in fact, no need to introduce the "arbitrary" primitive root of unity ($\omega = \epsilon^k$); it would be enough to carry out all the calculations only for $\omega = \epsilon$, concluding that ϵ is a root of the polynomial $g(z)$ defined by (7).

Now, every polynomial with integer coefficients that has ϵ as a root is divisible by $h(z) = 1 + z + \cdots + z^{p-1}$ (the *minimal polynomial* of ϵ; this is a known fact of algebra); hence the equality $g(z) = c \cdot h(z)$, i.e., (8). This is exactly the way in which the paper was presented by IMO contestant Nikolay Nikolov (from the Bulgarian team); his solution was awarded a special prize.

Thirty-seventh International Olympiad, 1996

1996/1

Assume that AB is the lower horizontal base of the rectangle $ABCD$. If the coin is moved from the square with center P to the square with center Q, this move is represented by the vector $\overrightarrow{PQ} = (x, y)$ with integer entries x, y satisfying the equation $x^2 + y^2 = r$.

(a) When r is even, integers x, y satisfying this equation are both even or odd. Color the squares of the board in chessboard pattern; the coin moves over squares of the same color. The corner squares at vertices A and B have different colors, so they cannot be linked by a sequence of moves.

When r is divisible by 3, the equation $x^2 + y^2 = r$ implies that x and y are both divisible by 3 (because $x^2 \equiv 1 \pmod 3$ when $x \not\equiv 0 \pmod 3$). We color black every third vertical row, starting from the row adjacent to AD. The coin that starts from the square with vertex A keeps moving over black squares and cannot reach the square with vertex B.

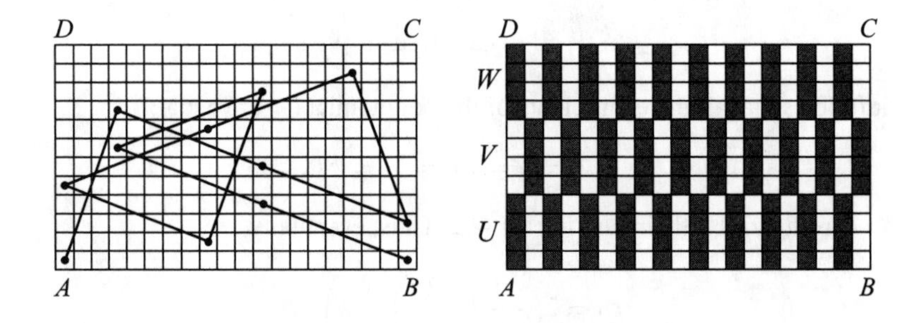

(b) For $r = 73 = 8^2 + 3^2$, legal moves are translations through vectors $(\pm 8, \pm 3)$ and $(\pm 3, \pm 8)$. An example of a sequence of eleven moves connecting the two lower corner squares is illustrated in the diagram.

(c) The only representation of $r = 97$ as a sum of two square numbers is $97 = 9^2 + 4^2$; translations by the vectors $(\pm 9, \pm 4)$ and $(\pm 4, \pm 9)$ are now the only legal moves. Partition $ABCD$ into three equal (20×4)-rectangles U, V, W (U is the lower strip, V is the middle one, and W is the upper strip). In each one of these three regions let us color the vertical rows ((1×4)-rectangles) alternately black and white, in "zebra" pattern. In regions U and W the leftmost vertical row is colored black, and in V the leftmost row is colored white.

We claim that every move from a black square goes to a black square. Indeed: a move of type $(\pm 9, \pm 4)$ always connects two squares, one of which is in the set $U \cup W$ and the other in V, and the horizontal component of the vector is an odd integer, while a move of type $(\pm 4, \pm 9)$ can only connect two squares, one in U, the other in W, the horizontal component of the move vector being an even integer.

The coin starts from the black square at vertex A; every move takes it into another black square, so it will never reach the white square at vertex B. Hence the answer: for $r = 97$ the task cannot be done.

1996/2

First Solution. Line BD (bisector of angle B in triangle ABP) intersects segment AP at a point X which divides that segment in the ratio $AX/PX = AB/PB$. Analogously, CE cuts the same segment at a point Y dividing it in the ratio $AY/PY = AC/PC$. To show that X and Y coincide (and determine the point of concurrence of lines AP, BD and CE), it is enough to show that these ratios are equal:

$$\frac{AB}{PB} = \frac{AC}{PC}. \tag{1}$$

On the ray symmetric to AP with respect to the bisector of angle CAB lay off AQ so that

$$\frac{AQ}{AC} = \frac{AB}{AP}. \tag{2}$$

From the definition of the ray AQ follow the equalities $\angle PAB = \angle QAC$ and $\angle PAC = \angle QAB$, which together with (2) show that triangle APB is similar to ACQ and triangle APC is similar to ABQ. Hence

$$\angle APB = \angle ACQ, \qquad \angle APC = \angle ABQ, \tag{3}$$

and

$$\frac{AB}{PB} = \frac{AQ}{CQ}, \qquad \frac{AC}{PC} = \frac{AQ}{BQ}. \tag{4}$$

In view of formulas (3), the angle equality from the condition of the problem becomes simply $\angle BCQ = \angle CBQ$. So the triangle BCQ is isosceles, $BQ = CQ$. Thus the right sides of equalities (4) are equal; their left sides must be equal, too, and we have the claimed equality (1).

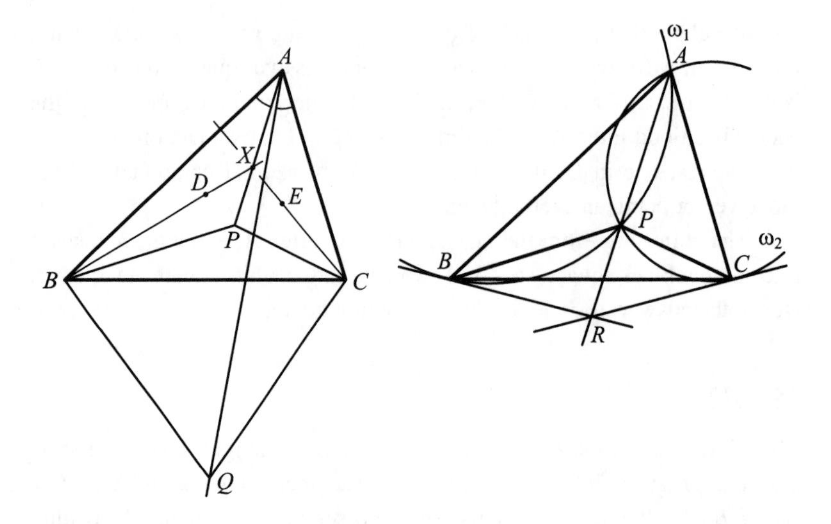

Second Solution. Let ω_1 and ω_2 be the circumcircles of triangles APB and APC. The tangents to these circles at B and C intersect in a point R so that $\angle APB = 180^\circ - \angle ABR$ and $\angle APC = 180^\circ - \angle ACR$, by the tangent-chord theorem for the chords AB and AC. Inserting these relations into the equation from the problem statement we get

$$\angle ABR - \angle ABC = \angle ACR - \angle ACB;$$

equivalently: $\pm\angle CBR = \pm\angle BCR$ (on both sides minus or plus sign, depending on whether R and A lie on the same side or on opposite sides of line BC).

In either case, BCR is an isosceles triangle: $BR = CR$. Thus R has equal powers ($BR^2 = CR^2$) with respect to the circles ω_1 and ω_2, and consequently lies on the radical axis of this pair of circles, i.e., on line AP (see *Glossary*).

The power of R with respect to each one of these circles is also equal to the product $RA \cdot RP$. We get the equalities

$$\frac{RP}{RB} = \frac{RB}{RA} \quad \text{and} \quad \frac{RP}{RC} = \frac{RC}{RA},$$

which show that triangle RPB is similar to RBA (notice the common angle R) and triangle RPC is similar to RCA (analogously). Therefore

$$\frac{RB}{PB} = \frac{RA}{BA} \quad \text{and} \quad \frac{RC}{PC} = \frac{RA}{CA};$$

and since $RB = RC$, we obtain equality (1) from the first solution, which suffices to deduce the assertion of the problem.

Third Solution. Extend segments AP, BP, CP to their second intersections with the circumcircle of triangle ABC at the respective points M, K, L. Then

$$\angle APB = \angle KPM = \angle PAK + \angle AKP = \angle MAK + \angle ACB,$$
$$\angle APC = \angle LPM = \angle PAL + \angle ALP = \angle MAL + \angle ABC.$$

The equality from the condition of the problem says $\angle MAK = \angle MAL$. So $MK = ML$.

As in the two previous solutions, we again reduce the proof to equality (1). Triangle PKM is similar to PAB, and triangle PLM is similar to PAC (the corresponding angles are equal). Hence

$$\frac{KM}{PM} = \frac{AB}{PB} \quad \text{and} \quad \frac{LM}{PM} = \frac{AC}{PC}.$$

In view of $KM = LM$, equality (1) follows.

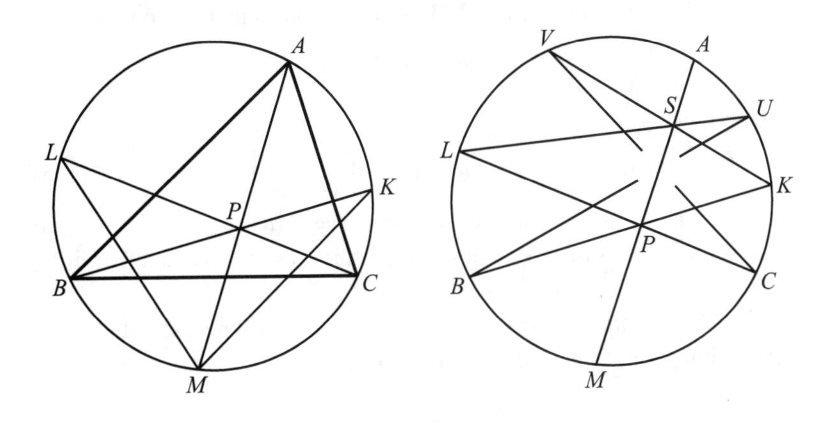

Fourth Solution. In this solution we do *not* reduce the problem to a proof of equality (1). As in the third solution, we consider the circumcircle of triangle ABC and derive equality $MK = ML$. Let U and V be the midpoints of arcs AK and AL that do not contain M. Rays KV and LU are the bisectors of angles K and L in triangle AKL, so they intersect at a point S lying on the third angle bisector, i.e., on ray AM.

In the hexagon $BCKUVL$, the diagonals KB and LC meet at P, and the diagonals KV and LU meet at S. By virtue of Pascal's theorem, the third pair of diagonals (BU and CV) meet at a point on line PS (see *Glossary*); i.e., on line AP. It remains to note that BU and CV are the bisectors of angles ABP and ACP, and we have the result.

1996/3

Suppose a function $f\colon \mathrm{N}_0 \to \mathrm{N}_0$ satisfies the given equation:

$$f(m + f(n)) = f(f(m)) + f(n) \quad \text{for } m, n \in \mathrm{N}_0. \tag{1}$$

Setting $m = n = 0$ we get $f(0) = 0$. Let V be the set of all values taken by f. Setting in (1) $m = 0$ and denoting $f(n)$ by v we obtain

$$f(v) = v \quad \text{for } v \in V. \tag{2}$$

Equation (1) becomes:

$$f(m + v) = f(m) + v \quad \text{for } m \in \mathrm{N}_0, v \in V. \tag{3}$$

The function identically equal to zero is of course a solution. For the sequel assume that f takes also nonzero values. Let d be the least positive number in the set V; then $f(d) = d$ by (2). Easy induction using (3) shows that

$$kd \in V \quad \text{for } k \in \mathrm{N}_0. \tag{4}$$

We claim that

$$\text{if } u, v \in V, \quad u > v, \quad \text{then} \quad u - v \geq d. \tag{5}$$

Indeed: taking in (3) $m = u - v$ we obtain $u = f(u) = f(u - v) + v$, so that $u - v = f(u - v) \in V$; and since d is the least positive number in V, we see that $u - v \geq d$, as claimed.

It follows from (4) and (5) that the set V consists precisely of all non-negative integer multiples of d. Thus the formula

$$h(r) = \frac{f(r)}{d} \quad \text{for } r = 0, 1, \ldots, d - 1 \tag{6}$$

defines a function

$$h\colon \{0, \ldots, d - 1\} \to \mathrm{N}_0 \quad \text{with} \quad h(0) = 0. \tag{7}$$

Every integer $n \in \mathrm{N}_0$ is uniquely representable in the form

$$n = qd + r \quad \text{where} \quad q \in \mathrm{N}_0, \quad r \in \{0, \ldots, d - 1\}; \tag{8}$$

for n, q, r as in (8) we have in view of (3) (with $m = r$, $v = qd$):

$$f(n) = f(r) + qd = (q + h(r))d. \tag{9}$$

We are going to show that also, conversely, if d is a fixed positive integer and h is any function of type (7), then the formula (9) (for n as in (8)) defines a function $f\colon \mathrm{N}_0 \to \mathrm{N}_0$ satisfying equation (1).

The condition $h(0) = 0$ ensures that if v is an integer divisible by d (so that $v = qd + 0$ in (8)) then, by (9), $f(v) = qd = v$.

Let m and n be any numbers in N_0; consider their representations

$$m = kd + s, \qquad n = \ell d + t, \qquad k, \ell \in \mathrm{N}_0, \quad s, t \in \{0, \dots, d-1\}.$$

According to the definition (9),

$$f(m) = (k + h(s))d, \qquad f(n) = (\ell + h(t))d;$$

the number $v = f(m)$ is divisible by d, and therefore $f(v) = v$; equivalently, $f(f(m)) = (k + h(s))d$.

Note that $m + f(n) = (k + \ell + h(t))d + s$. Applying once more definition (9) we obtain

$$f(m + f(n)) = (k + \ell + h(t) + h(s))d = f(n) + f(f(m));$$

equation (1) is fulfilled.

Summarizing, there exist infinitely many functions $f\colon \mathrm{N}_0 \to \mathrm{N}_0$ satisfying equation (1) and not equal identically to zero; the general form of such a function is determined by an arbitrary choice of a positive integer d and an arbitrary choice of a function $h\colon \{0, \dots, d-1\} \to \mathrm{N}_0$ such that $h(0) = 0$; the value $f(n)$ for an n as in (8) is then defined by formula (9).

1996/4

According to hypothesis,

$$15a + 16b = u^2 \quad \text{and} \quad 16a - 15b = v^2 \tag{1}$$

for some $u, v \geq 1$. Hence $u^4 + v^4 = (15^2 + 16^2)(a^2 + b^2) = 13 \cdot 37(a^2 + b^2)$ and we obtain the system of two congruences

$$u^4 + v^4 \equiv 0 \pmod{13} \quad \text{and} \quad u^4 + v^4 \equiv 0 \pmod{37}. \tag{2}$$

In each of them, either both summands are congruent to zero, or none. And since 13 and 37 are prime numbers, we get that u and v are both divisible or nondivisible by 13 and are both divisible or nondivisible by 37.

Now, using the identities $u^{12}+v^{12}=(u^4+v^4)(u^8-u^4v^4+v^8)$ and $u^{36}+v^{36}=(u^{12}+v^{12})(u^{24}-u^{12}v^{12}+v^{24})$ we infer from (2) that

$$u^{12}+v^{12}\equiv 0 \pmod{13} \quad \text{and} \quad u^{36}+v^{36}\equiv 0 \pmod{37}. \tag{3}$$

If u and v were nondivisible by 13, then by Fermat's little theorem (see *Glossary*) both summands in the first relation of (3) would be congruent to 1 (mod 13), a contradiction; and if u and v were nondivisible by 37, then both summands in the second relation of (3) would be congruent to 1 (mod 37); a contradiction again.

Thus u and v are divisible by 13 and by 37; hence they are divisible by $13\cdot 37=481$, and the square of each of them is divisible by 481^2. Therefore $\min\{u^2,v^2\}\geq 481^2$. On the other hand, for $u=v=481$ formulas (1) constitute a system of linear equations (with unknowns a and b), with the solution $a=481\cdot 31, b=481$. Thus 481^2 is the minimum value we are looking for.

Remark. Instead of resorting to Fermat's little theorem, one can simply inspect all possible residues of x^4 in division by 13 and by 37 and verify directly (an unpleasant task) that each one of the congruences (2) has only the trivial solution: $u\equiv v\equiv 0 \pmod{13}$ and $u\equiv v\equiv 0 \pmod{37}$.

1996/5

First Solution. Let $AB=a$, $BC=b$, $CD=c$, $DE=d$, $EF=e$, $FA=f$. It follows from the conditions of the problem that the internal angles of the hexagon satisfy $\angle A=\angle D$, $\angle B=\angle E$, $\angle C=\angle F$.

Choose a pair of parallel sides, say, BC and EF; denote by P and Q the feet of perpendiculars from A and D to line BC, and by S and T the feet of perpendiculars from D and A to line EF. Then

$$FB\geq TP=TA+AP=FA\cdot\sin F+AB\cdot\sin B=f\sin C+a\sin E,$$

and also

$$FB\geq QS=QD+DS=CD\cdot\sin C+DE\cdot\sin E=c\sin C+d\sin E;$$

adding these two inequalities, $2\cdot FB\geq(c+f)\sin C+(a+d)\sin E$. By the law of sines (for triangle FAB), $FB=2R_A\sin A$; thus we get

$$4R_A\geq(c+f)\frac{\sin C}{\sin A}+(a+d)\frac{\sin E}{\sin A}.$$

Considering other pairs of parallel sides ($AB \parallel DE$, $FA \parallel CD$) we obtain analogously

$$4R_C \geq (e+b)\frac{\sin E}{\sin C} + (c+f)\frac{\sin A}{\sin C},$$
$$4R_E \geq (a+d)\frac{\sin A}{\sin E} + (e+b)\frac{\sin C}{\sin E}.$$

It is now enough to add these three inequalities to obtain the required one:

$$\begin{aligned} 4(R_A + R_C + R_E) &\geq (a+d)\left(\frac{\sin E}{\sin A} + \frac{\sin A}{\sin E}\right) \\ &\quad + (b+e)\left(\frac{\sin C}{\sin E} + \frac{\sin E}{\sin C}\right) \\ &\quad + (c+f)\left(\frac{\sin A}{\sin C} + \frac{\sin C}{\sin A}\right) \\ &\geq 2(a+d) + 2(b+e) + 2(c+f) = 2p. \end{aligned}$$

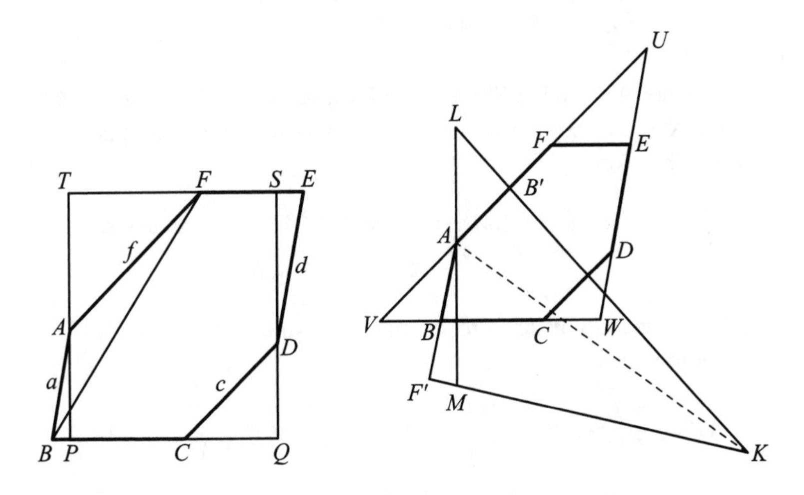

Second Solution. Lines BC, DE, FA intersect pairwise at points U, V, W so that the segments BC, DE, FA are contained respectively in the sides VW, WU, UV of the triangle UVW. Denote the lengths of these sides by u, v, w, respectively. Moreover, let

$$x = \frac{WE}{WU} = \frac{FV}{UV}, \quad y = \frac{UA}{UV} = \frac{BW}{VW}, \quad z = \frac{VC}{VW} = \frac{DU}{WU};$$

these are numbers from the interval (0, 1). Then

$$EF = (1 - x)u, \quad AB = (1 - y)v, \quad CD = (1 - z)w. \tag{1}$$

From the definition of x, y, and z we also have

$$BC = VC + BW - VW = VW(z + y - 1) = (y + z - 1)u, \tag{2}$$

and likewise,

$$DE = (z + x - 1)v, \quad FA = (x + y - 1)w. \tag{3}$$

Summing all the equalities in (1), (2), (3) we express the perimeter of the hexagon as follows:

$$p = x(v + w - u) + y(w + u - v) + z(u + v - w). \tag{4}$$

Consider triangle FAB. On rays AF and AB lay off segments $AB' = AB$ and $AF' = AF$. Through the points A, F' and B' draw lines perpendicular (respectively) to VW, WU, and UV. These lines intersect in pairs at points K, L, M so that $LM \perp VW$, $MK \perp WU$, $KL \perp UV$. Triangle KLM is similar to UVW.

Segments AB' and AF' are altitudes in triangles KAL and KAM; segment AK is not shorter than the altitude in triangle KLM dropped from vertex K. Therefore

$$\begin{aligned} AK \cdot LM \geq 2 \cdot \text{area}(\Delta KLM) &= 2 \cdot \text{area}(\Delta KAL) + 2 \cdot \text{area}(\Delta KAM) \\ &= KL \cdot AB' + MK \cdot AF'. \end{aligned}$$

Since KLM and UVW are similar triangles, while $AB'F'$ and ABF are congruent triangles,

$$AK \geq \frac{KL}{LM} \cdot AB' + \frac{MK}{LM} \cdot AF' = \frac{UV}{VW} \cdot AB + \frac{WU}{VW} \cdot FA. \tag{5}$$

Note that AK is the diameter of the circumcircle of triangle $AB'F'$, congruent to ABF; so $AK = 2R_A$. Inserting (1), (2), (3) into (5) we get

$$2R_A \geq \frac{w}{u}(1 - y)v + \frac{v}{u}(x + y - 1)w = x \cdot \frac{vw}{u},$$

and by analogy,

$$2R_C \geq y \cdot \frac{wu}{v}, \qquad 2R_E \geq z \cdot \frac{uv}{w}.$$

Adding these three inequalities and applying formula (4) we obtain

$$\begin{aligned} 2(R_A + R_C + R_E) - p &\geq x \cdot \frac{vw}{u} + y \cdot \frac{wu}{v} + z \cdot \frac{uv}{w} - p \\ &= x\left(\frac{vw}{u} - v - w + u\right) + y\left(\frac{wu}{v} - w - u + v\right) \\ &\quad + z\left(\frac{uv}{w} - u - v + w\right). \qquad (6) \end{aligned}$$

The numbers x, y, z satisfy the inequalities $y + z > 1 > x$, $z + x > y$, $x + y > z$, as can be seen from (2), (3). So there exist positive numbers x', y', z' such that $x = y' + z'$, $y = z' + x'$, $z = x' + y'$. Inserting these into (6) and grouping together the terms containing x', y' and z' we transform the right-hand expression of (6) into the following:

$$ux'\left(\frac{v}{w} + \frac{w}{v} - 2\right) + vy'\left(\frac{w}{u} + \frac{u}{w} - 2\right) + wz'\left(\frac{u}{v} + \frac{v}{u} - 2\right).$$

The value of this last expression is obviously nonnegative; hence the result.

Remark. Let XYZ be any triangle, let P be any point inside it and let B, D, F be the orthogonal projections of P onto lines XY, YZ, ZX. Complete the parallelograms $PFAB$, $PBCD$, $PDEF$; a convex hexagon $ABCDEF$ appears with opposite sides pairwise parallel and equal. The perimeter of the hexagon equals $2 \cdot PB + 2 \cdot PD + 2 \cdot PF$.

Let R_A, R_B, R_C be the circumradii of triangles FAB, BCD, DEF, as in the problem. The segment PX is the diameter of the circumcircle of triangle BPF, which is congruent to FAB; hence $PX = 2R_A$, and similarly, $PY = 2R_C$, $PZ = 2R_E$. According to the assertion of the problem, the sum of these three numbers is not less than the perimeter of the hexagon. Thus

$$PX + PY + PZ \geq 2(PB + PD + PF).$$

In words: the sum of the distances from any point inside a triangle to the vertices of the triangle is not less than twice the sum of the distances from that point to the sides. This is a classical theorem, known as the *Erdős–Mordell inequality*. We have derived it as a corollary of the result of the IMO problem under discussion, applied to a hexagon in which the opposite sides are not only parallel but also have equal lengths. This last condition is not present in the problem statement. Therefore the problem can be considered as an essential strengthening of the Erdős–Mordell inequality.

1996/6

Assume that the difference $x_i - x_{i-1}$ equals p for k values of i and equals $-q$ for m values of i. Then $k + m = n$ and

$$0 = x_n - x_0 = \sum_{i=1}^{n}(x_i - x_{i-1}) = kp - mq. \tag{1}$$

Let d be the greatest common divisor of k and m, and δ be the greatest common divisor of p and q. Hence $k = ad$, $m = bd$, $q = \alpha\delta$, $p = \beta\delta$. Substituting these products into the equality $kp = mq$ (see (1)) we get $a\beta = b\alpha$. Since a and b are coprime, and also α and β are coprime, it follows that $a = \alpha$, $b = \beta$. Thus $k = ad$, $m = bd$, $p = b\delta$, $q = a\delta$.

We introduce further notation: $c = a + b$, $s = p + q$, so that $cd = n$, $c\delta = s$, and we set $r_i = x_i - x_{i-c}$ for $i = c, c+1, \ldots, n$. The numbers r_i thus defined satisfy the following relations:

$$r_c + r_{2c} + \cdots + r_{dc} = x_{dc} - x_0 = x_n - x_0 = 0 \tag{2}$$

and (for $i = c, c+1, \ldots, n-1$)

$$r_{i+1} - r_i = (x_{i+1} - x_i) - (x_{i-c+1} - x_{i-c}).$$

Each one of the two differences in the parentheses has value p or $-q$, according to the conditions of the problem. And since $p + q = s$, we see that

$$r_{i+1} - r_i \in \{0, s, -s\} \quad \text{for } i = c, c+1, \ldots, n-1. \tag{3}$$

The number r_c can be written in the form

$$r_c = x_c - x_0 = (x_1 - x_0) + (x_2 - x_1) + \cdots + (x_c - x_{c-1});$$

this is the sum of c summands, each one equal to p or $-q$. If ℓ summands have value p and the remaining $c - \ell$ summands have value $-q$, we get that the whole sum equals

$$r_c = \ell p - (c - \ell)q = \ell s - cq = \ell s - ca\delta = (\ell - a)s.$$

Hence, in view of (3), we infer that all r_is are divisible by s. Therefore the sequence $r_c, r_{c+1}, r_{c+2}, \ldots, r_{n-1}, r_n$ is composed of multiples of s. Every two consecutive terms in this sequence differ (in absolute value) at most by s, also in view of (3). Assuming that all terms are different from zero, we conclude that they must be all positive or all negative. In particular, the numbers $r_c, r_{2c}, \ldots, r_{dc}$ must have the same sign. However, this is not possible, as can be seen from (2).

Thus the equality $r_j = 0$ holds for a certain $j \in \{c, c+1, \ldots, n\}$; this means that $x_j = x_{j-c}$. And since $c \leq c\delta = s = p + q < n$, the pair of indices $(j - c, j)$ is not identical with $(0, n)$; so it is the pair whose existence had to be proved.

Thirty-eighth International Olympiad, 1997

1997/1

(a) Let ABC be one of the right triangles in question, with legs of lengths $AB = m$, $BC = n$. Complete the rectangle $ABCD$. If m and n are of the same parity, then the coloring pattern of $ABCD$ is centrosymmetric about the center of the rectangle. Thus the area of the black part of the triangle ABC (i.e., S_1) is an exact half of the "black area" of $ABCD$. The same can be said about the "white area" (S_2) of the triangle.

When m and n are even, then the rectangle $ABCD$ is exactly half-black and half-white; so $S_1 = S_2$, and hence $f(m, n) = 0$ in that case. When m and n are odd, the rectangle $ABCD$ contains one black square more or less than it contains white squares. In this case, the outcome is $f(m, n) = |S_1 - S_2| = 1/2$.

(b) For m and n of the same parity, the value of $f(m, n)$ is 0 or 1/2, and there is nothing to prove. We now inspect the remaining case. Assume that e.g., m is even and n is odd. On side CB lay off $CE = 1$. Triangle

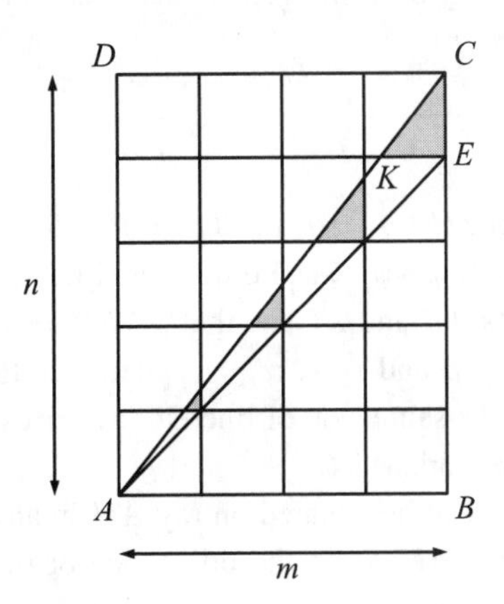

ABE has legs of even lengths, so it is half-black, half-white, according to the conclusion of part (a). Thus the difference between the black area and the white area of triangle ABC is the same as the analogous difference for triangle AEC, and cannot exceed the total area of the latter, equal to $m/2$. Assuming that n is even and m is odd, we analogously obtain the upper estimate $n/2$. The claim follows: $f(m,n) \le \frac{1}{2}\max\{m,n\}$.

(c) Now consider a rectangle $ABCD$ with sidelengths $AB = m$ and $BC = n = m+1$, where m is an even integer. Assume the unit square at vertex A is white. Define E as in (b) ($CE = 1$). Let K be the point on AC such that $KE \parallel AB$.

The black part of triangle AEC consists of m small triangles: KEC and $m-1$ similar copies of KEC, in ratios $(m-1)/m$, $(m-2)/m, \ldots, 2/m$, $1/m$. The legs of KEC have lengths $KE = m/(m+1)$, $EC = 1$. Therefore

$$\begin{aligned}\text{black area}(\triangle AEC) &= \left(\left(\frac{1}{m}\right)^2 + \left(\frac{2}{m}\right)^2 + \cdots + \left(\frac{m}{m}\right)^2\right) \cdot \text{area}(\triangle KEC)\\ &= \frac{m(m+1)(2m+1)}{6m^2} \cdot \frac{m}{2(m+1)} = \frac{2m+1}{12};\\ \text{white area}(\triangle AEC) &= \text{total area}(\triangle AEC) - \text{black area}(\triangle AEC)\\ &= \frac{m}{2} - \frac{2m+1}{12} = \frac{4m-1}{12}.\end{aligned}$$

In the proof of (b) we have observed that $f(m,n)$ is the difference between these two areas. Thus, in the case under consideration,

$$f(m, m+1) = \left|\frac{2m+1}{12} - \frac{4m-1}{12}\right| = \frac{m-1}{6}.$$

And since m can be any even integer, we see that there is no upper bound for the values of $f(m,n)$.

1997/2

First Solution. Let $\angle CAB = \alpha$, $\angle ABC = \beta$, $\angle BCA = \gamma$, $\angle BAU = \varphi$, $\angle CAU = \psi$; obviously $\varphi + \psi = \alpha$. Points V and W lie on the perpendicular bisectors of segments AB and AC, so that $\angle ABV = \varphi$, $\angle ACW = \psi$. The inequalities $\varphi < \alpha \le \beta$ and $\psi < \alpha \le \gamma$ guarantee that the rays BV and CW are situated on the same side of line BC as vertex A; so the points V, W and T lie inside triangle ABC.

Points V and W can be situated on ray AU in any order; point T can lie on one side of line AU or on the other; one of the angles β and γ is certainly acute; the other one can (but need not) be so. Nevertheless, the reasoning that follows is case-independent.

Extend CT to the intersection with the circumcircle of triangle ABC at X (on the arc AB which does not contain C). Triangle AWC is isosceles; so the lines AW and CW lie symmetrically with respect to the perpendicular bisector of AC, which is also the axis of symmetry of the circumcircle. Consequently, AU and CX are symmetric chords of that circle. So $AU = CX$.

We have $\angle TBX = \angle TBA + \angle ABX = \angle TBA + \angle ACX = \varphi + \psi = \alpha$. This equality shows that BTX is an isosceles triangle: $TB = TX$. Thus, finally, $TB + TC = TX + TC = CX = AU$.

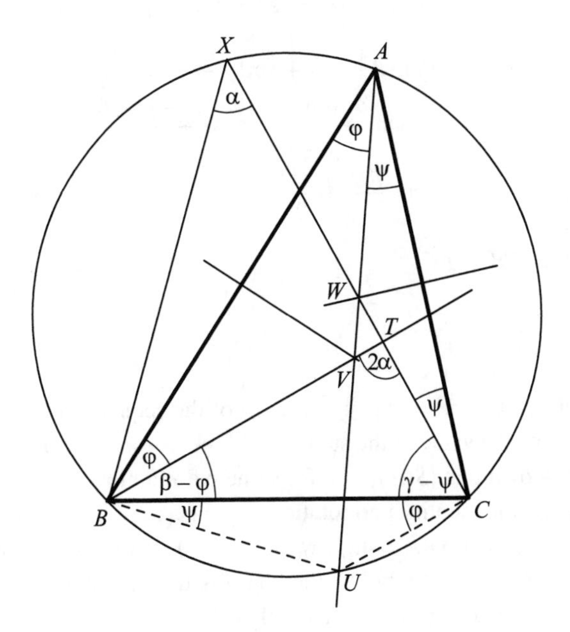

Second Solution. We define $\alpha, \beta, \gamma, \varphi, \psi$ as in the first solution and conclude (as before) that V, W, T lie inside triangle ABC. Also this solution is case-independent. In triangle TBC,

$$\angle TBC + \angle TCB = (\beta - \varphi) + (\gamma - \psi) = \beta + \gamma - \alpha = 180^\circ - 2\alpha,$$

and hence $\angle BTC = 2\alpha$. Apply the law of sines to triangles BCT and ABC:

$$\frac{TB}{\sin(\gamma-\psi)} = \frac{TC}{\sin(\beta-\varphi)} = \frac{BC}{\sin 2\alpha} = \frac{2R\sin\alpha}{\sin 2\alpha} = \frac{R}{\cos\alpha},$$

where R is the circumradius of ABC. Thus

$$\begin{aligned} TB+TC &= \frac{R}{\cos\alpha}(\sin(\beta-\varphi)+\sin(\gamma-\psi)) \\ &= \frac{R}{\cos\alpha}\cdot 2\sin\frac{\beta+\gamma-\alpha}{2}\cos\frac{\beta-\gamma-\varphi+\psi}{2} \\ &= 2R\cos\frac{\beta-\gamma-\varphi+\psi}{2}. \end{aligned}$$

On the other hand, from the law of sines for triangles ABU and ACU we get $AU = 2R\sin(\beta+\psi)$ and $AU = 2R\sin(\gamma+\varphi)$, so that

$$\begin{aligned} AU &= R(\sin(\beta+\psi)+\sin(\gamma+\varphi)) \\ &= R\cdot 2\sin\frac{\alpha+\beta+\gamma}{2}\cos\frac{\beta+\psi-\gamma-\varphi}{2} \\ &= 2R\cos\frac{\beta-\gamma-\varphi+\psi}{2}. \end{aligned}$$

Hence $AU = TB+TC$.

1997/3

For any permutation $\pi = (y_1, y_2, \ldots, y_n)$ of the sequence $(x_1, x_2, \ldots, x_n)$ let $S(\pi)$ denote the value of the sum $y_1+2y_2+\cdots+ny_n$. Let $r = (n+1)/2$. It has to be shown that $|S(\pi)| \le r$ for some permutation π.

Let π_0 be the identity permutation $\pi_0 = (x_1, x_2, \ldots, x_n)$ and let $\tilde{\pi}$ be the reverse permutation $\tilde{\pi} = (x_n, x_{n-1}, \ldots, x_1)$. If one of the inequalities $|S(\pi_0)| \le r$ or $|S(\tilde{\pi})| \le r$ holds, the claim is true. Assume for the sequel that $|S(\pi_0)| > r$ and $|S(\tilde{\pi})| > r$. Note that

$$S(\pi_0)+S(\tilde{\pi}) = (n+1)(x_1+x_2+\cdots+x_n),$$

and hence $|S(\pi_0)+S(\tilde{\pi})| = n+1 = 2r$. Since each one of the numbers $S(\pi_0)$ and $S(\tilde{\pi})$ exceeds r in absolute value, they must have opposite signs. Accordingly, one of them is greater than r and the other one is smaller than $-r$.

Starting from π_0, we are able to obtain any permutation by successive transpositions of neighboring elements. In particular, there exists a chain of permutations $\pi_0, \pi_1, \ldots, \pi_m$ such that $\pi_m = \tilde{\pi}$ and, for each $i \in \{0, \ldots, m-1\}$, the permutation π_{i+1} arises from π_i by interchanging two of its neighboring terms. This means that if $\pi_i = (y_1, y_2, \ldots, y_n)$ and $\pi_{i+1} = (z_1, z_2, \ldots, z_n)$, then there is an index $k \in \{1, \ldots, n-1\}$ such that

$$z_k = y_{k+1}, \qquad z_{k+1} = y_k; \qquad z_j = y_j \quad \text{for } j \neq k, k+1.$$

And since the numbers x_i do not exceed r in absolute value,

$$\begin{aligned} |S(\pi_{i+1}) - S(\pi_i)| &= |kz_k + (k+1)z_{k+1} - ky_k - (k+1)y_{k+1}| \\ &= |y_k - y_{k+1}| \leq |y_k| + |y_{k+1}| \leq 2r. \end{aligned}$$

This shows that the distance between any two successive numbers in the sequence $S(\pi_0), S(\pi_1), \ldots, S(\pi_m)$ is not greater than $2r$.

Recall that the numbers $S(\pi_0)$ and $S(\pi_m)$, regarded as points of the real line, lie outside the interval $[-r, r]$, on distinct sides of it. It follows that at least one of the numbers $S(\pi_i)$ has to hit that interval. Thus we have $|S(\pi_i)| \leq r$ for a certain π_i, and the result is proved.

1997/4

First Solution. (a) Let A be an $n \times n$ silver matrix ($n > 1$). We will show that n is an even integer (hence different from 1997).

Let a_{ij} be the number at the intersection of the ith row and the jth column. Choose and fix an element $c \in S$ that does not appear on the main diagonal of A (i.e., $c \neq a_{ii}$ for all i). Two indices $i, j \in \{1, \ldots, n\}$ will be called *conjugate* if either $a_{ij} = c$ or $a_{ji} = c$. Take any index i. Consider the "cross" composed of all the entries in the ith row and the ith column. By the definition of a silver matrix, every element of S occurs in this cross, each of them exactly once (as there are only $2n - 1$ cells in the cross). In particular, c appears exactly once as a certain a_{ij} or a_{ji}, where $j \neq i$.

Thus every index $i \in \{1, \ldots, n\}$ is conjugated with exactly one index $j \in \{1, \ldots, n\}, j \neq i$. So the elements of the set $\{1, \ldots, n\}$ are matched into pairs of mutual conjugates. Consequently, n is even.

(b) We now show how to create a $2n \times 2n$ silver matrix from an $n \times n$ silver matrix.

Let M be any n-element set. It is not hard to construct a matrix whose every row and every column is a permutation of the elements of M (such matrices are called *Latin squares*). Here is an example (by far, not unique):

for $M = \{x_1, \ldots, x_n\}$ the matrix

$$\begin{bmatrix} x_1 & x_2 & \ldots & x_{n-1} & x_n \\ x_2 & x_3 & \ldots & x_n & x_1 \\ \vdots & \vdots & \ddots & \vdots & \vdots \\ x_n & x_1 & \ldots & x_{n-2} & x_{n-1} \end{bmatrix},$$

is a Latin square.

Let A be a silver matrix with entries from the set $\{1, 2, \ldots, 2n - 1\}$. Let B be any $n \times n$ Latin square, with entries from the set $\{2n, \ldots, 3n - 1\}$, and C be an $n \times n$ Latin square, with entries from the set $\{3n, \ldots, 4n - 1\}$. Consider the $2n \times 2n$ matrix

$$\begin{bmatrix} A & B \\ C & A \end{bmatrix} \quad \text{(block form).}$$

For each $i \in \{1, 2, \ldots, 2n\}$, the ith row and the ith column of this new matrix make up a "cross" composed of three parts: a certain "cross" of the matrix A, a line (row or column) of B and a line of C. These three parts jointly contain all the elements of each one of the sets $\{1, 2, \ldots, 2n - 1\}$, $\{2n, \ldots, 3n - 1\}$, $\{3n, \ldots, 4n - 1\}$. So we have defined a $2n \times 2n$ silver matrix. Thus, starting from the 1×1 silver matrix $[1]$, we can inductively obtain $n \times n$ silver matrices for all n that are integral powers of 2.

Second Solution. Part (a) is proved as in the first solution. To prove (b), fix a positive even integer n and consider a tournament of n players, each two playing one game. It is a well-known fact (proofs of which can be found in most books on graph theory) that such a "round-robin" tournament for n even can be organized so that the games are played during $n-1$ consecutive days, each player playing exactly one game each day.

Label the players 1 through n; assume the tournament has been played twice and number the consecutive days $1, \ldots, n - 1$ (the first "complete round" of the tournament) and $n, \ldots, 2n-2$ (the second "complete round"). We define an $n \times n$ matrix $[a_{ij}]$ as follows.

For any two indices $i < j$ let a_{ij} be the number of the day of the first game between players i and j, and let a_{ji} be the number of the day of the second game between those players. For each $i \in \{1, \ldots, n\}$, look at the "cross" composed of all the entries in the ith row and the ith column, without the cell (i, i); it contains all the numbers from 1 to $2n - 2$. It now suffices to write the number $2n - 1$ everywhere on the main diagonal, to obtain a silver matrix. In conclusion, an $n \times n$ silver matrix exists for *every* even integer $n \geq 2$.

1997/5

Take any pair of integers $a, b \geq 1$ satisfying the equation. We will consider two cases.

If $a \leq b$, we rewrite the equation as $(a^b)^b = b^a$. The exponents satisfy the inequality $b \geq a$, so the bases of exponentiation must satisfy the opposite inequality $a^b \leq b$. The number a cannot be greater than 1 because $2^b > b$ for every integer $b \geq 1$, as can be easily proved by induction. Hence $a = 1$. We see from the equation that also $b = 1$. Obviously, the pair $(a, b) = (1, 1)$ is a solution.

It remains to consider the case where $a > b$. Again we look at the original equation. Comparing the bases and exponents of the powers we infer that $b^2 < a$. Dividing the equation by b^{2b^2} we get

$$\left(ab^{-2}\right)^{b^2} = b^{a-2b^2}. \tag{1}$$

The left-hand expression has value greater than 1; thus also the right-hand expression must have value greater than 1. This implies the inequalities $a > 2b^2$ and $b > 1$.

Write the number a/b^2 as an irreducible fraction k/n. So $k = nab^{-2}$. Raising this equality to the power b^2 and using once more the given equation $a^{b^2} = b^a$ we obtain

$$k^{b^2} = n^{b^2}a^{b^2}b^{-2b^2} = n^{b^2}b^a b^{-2b^2} = n^{b^2}b^{a-2b^2}.$$

The exponent $a - 2b^2$ is positive, and hence both factors in the last expression are integers. Thus k^{b^2} is divisible by n; and since the fraction k/n is irreducible, $n = 1$.

So we have $a = kb^2$, with k being a positive integer. In fact, $k \geq 3$, as $a > 2b^2$. Insert $a = kb^2$ into (1) to obtain $k^{b^2} = b^{(k-2)b^2}$. Hence

$$k = b^{k-2}. \tag{2}$$

Since $b \geq 2$, the right side of (2) has value not less than 2^{k-2}. For any integer $k \geq 5$ we have the inequality $2^{k-2} > k$ (easy to prove by induction), so that equation (2) cannot be satisfied.

We are left with the values $k = 3$ and $k = 4$. Setting these values in (2) and keeping in mind that $a = kb^2$, we determine two pairs (a, b), which indeed satisfy the given equation: $(27, 3)$ and $(16, 2)$. Together with the pair $(1, 1)$ they constitute the complete solution of the equation.

1997/6

First Solution. Consider a representation of an odd integer $2k+1$ greater than 1. It contains at least one "1" (a summand of type 2^0). Deleting it we obtain a representation of $2k$. Conversely, adjoining a "1" to a representation of $2k$ we get a representation of $2k + 1$. These operations are mutually inverse. The first recurrence formula follows:

$$f(2k+1) = f(2k). \tag{1}$$

Now consider a representation of a positive even integer $2k$. If it contains at least one "1" then by deleting it we obtain a representation of $2k - 1$. As before, this is a one-to-one correspondence; so there are $f(2k-1)$ such representations. And if there is no "1" in the representation of $2k$ (considered) then we can divide all terms by 2 to obtain a representation of k. This is also a bijective correspondence; so there are $f(k)$ such representations. Thus the total number of representations of $2k$ is expressed by the second recurrence formula

$$f(2k) = f(2k-1) + f(k). \tag{2}$$

These formulas hold for $k = 1, 2, 3, \ldots$. Define $f(0) = 1$; equality (1) then holds for $k = 0$, as well. We see from (1) and (2) that f is nondecreasing.

According to (1), the number $f(2k-1)$ in (2) can be replaced by $f(2k-2)$:

$$f(2k) - f(2k-2) = f(k) \quad \text{for } k = 1, 2, 3, \ldots.$$

Fix an integer $r \geq 1$. Adding up these equalities for $k = 1, \ldots, r$ we obtain another useful recurrence formula

$$f(2r) = f(r) + f(r-1) + \cdots + f(2) + f(1) + f(0). \tag{3}$$

Since f is nondecreasing, $f(r)$ is the greatest term in this sum. If $r \geq 2$, then $f(r) \geq 2$, and so

$$\begin{aligned} f(2r) &= (f(r) + f(r-1) + \cdots + f(2)) + 1 + 1 \\ &\leq (r-1)f(r) + 2 \leq rf(r). \end{aligned}$$

Setting $r = 2^{n-1}, 2^{n-2}, \ldots, 2^2, 2^1$ we obtain an upper estimate of $f(2^n)$ (valid for any integer $n \geq 3$):

$$\begin{aligned} f(2^n) &\le 2^{n-1} f(2^{n-1}) \le 2^{n-1} \cdot 2^{n-2} f(2^{n-2}) \le \cdots \\ &\le 2^{n-1} \cdot 2^{n-2} \cdots 2^2 \cdot 2^1 f(2^1) = 2^{n(n-1)/2} \cdot 2 < 2^{n^2/2}; \end{aligned}$$

this is the right inequality from the assertion of the problem.

To work out the lower estimate, fix an even integer $r \ge 2$ and use formula (3), with $2r$ replacing r:

$$f(4r) = f(2r) + f(2r-1) + \cdots + f(0) = h(r) + \cdots + h(1) + 1,$$

where $h(i) = f(r+i) + f(r-i+1)$.

In view of (1) we have $h(1) = f(r+1) + f(r) = 2f(r)$; moreover, by (1) and (2),

$$\begin{aligned} h(i+1) - h(i) &= [f(r+i+1) - f(r+i)] - [f(r-i+1) - f(r-i)] \\ &= \begin{cases} 0 - 0 = 0 & \text{for } i \text{ even}, \\ f\left(\dfrac{r+i+1}{2}\right) - f\left(\dfrac{r-i+1}{2}\right) \ge 0 & \text{for } i \text{ odd}. \end{cases} \end{aligned}$$

Hence $h(r) \ge h(r-1) \ge \cdots \ge h(1)$, and we obtain the inequality

$$f(4r) \ge rh(1) + 1 > rh(1) = 2rf(r),$$

valid for every even $r \ge 2$. Note that it holds also for $r = 1$.

Take any integer $n \ge 2$ and let $\ell = \lfloor n/2 \rfloor$. In the last formula set $r = 2^{n-2}, 2^{n-4}, 2^{n-6}, \ldots, 2^{n-2\ell}$, to obtain

$$\begin{aligned} f(2^n) &> 2^{n-1} f(2^{n-2}) > 2^{n-1} \cdot 2^{n-3} f(2^{n-4}) > \cdots \\ &> 2^{n-1} \cdot 2^{n-3} \cdots 2^{n-2\ell+1} f(2^{n-2\ell}) = 2^{\ell n - \ell^2} f(2^{n-2\ell}) \\ &= \begin{cases} 2^{n^2/4} \cdot 1 = 2^{n^2/4} & \text{for } n \text{ even}, \\ 2^{(n^2-1)/4} \cdot 2 > 2^{n^2/4} & \text{for } n \text{ odd}, \end{cases} \end{aligned}$$

yielding the lower estimate as required.

Second Solution. The recursion formulas (1), (2), (3) and the upper estimate are proved as in the first solution. The lower estimate: from (3) we obtain, for every $q \ge 1$,

$$\begin{aligned} f(16q) &= f(8q) + \cdots + f(0) > f(8q) + \cdots + f(6q) > 2qf(6q) \\ &= 2q(f(3q) + \cdots + f(0)) \\ &> 2q(f(3q) + \cdots + f(2q)) > 2q \cdot qf(2q). \end{aligned}$$

Taking $q = 2^{n-1}$ we get the inequality

$$f(2^{n+3}) > 2^{2n-1} f(2^n) \quad \text{for } n = 1, 2, 3, \ldots, \tag{4}$$

enabling us to prove the claimed estimate $f(2^n) > 2^{n^2/4}$ by induction, with the inductive step from n to $n+3$ as follows.

Assume inductively that $f(2^n) > 2^{n^2/4}$ holds for a certain integer n. Then $f(2^{n+3}) > 2^{2n-1} \cdot 2^{n^2/4} = 2^{(n^2+8n-4)/4}$; and we need to obtain on the right side the number $2^{(n+3)^2/4}$. For this, the inequality $n^2 + 8n - 4 \geq (n+3)^2$ should be fulfilled; yet it fails to hold for $n \leq 6$. In other words, the induction step works only for $n \geq 7$.

To establish the induction base, we need to verify the claim for $n \leq 9$ directly. One can of course reach $f(512)$ within finite time using just the recursion formulas (1) and (2). A more efficient way to achieve that aim is the following: the values $f(8) = 10$, $f(16) = 36$, $f(32) = 202$ are calculated from (1) and (2). Then, by (3):

$$\begin{aligned} f(4j) &= [f(2j) + \cdots + f(j+1)] + [f(j) + \cdots + f(0)] \\ &> jf(j) + f(2j). \end{aligned}$$

Setting $j = 16$ and $j = 32$ we get $f(64) > 778 > 2^9$ and $f(128) > 7242 > 2^{12.25}$. Using (4) again, with $n = 5$ and $n = 6$, we get the desired bounds $f(256) > 2^9 \cdot 202 > 2^{16}$ and $f(512) > 2^{11} \cdot 778 > 2^{20.25}$, as needed.

Third Solution. If a representation of 2^n is composed of x_0 "ones," x_1 "twos," x_2 "fours," etc., x_n terms of type 2^n, then the following equation is satisfied:

$$x_0 + 2x_1 + 2^2 x_2 + \cdots + 2^n x_n = 2^n;$$

and the value $f(2^n)$ is equal to the number of solutions of this equation in nonnegative integers $x_0, x_1, \ldots, x_n$. Equivalently, $f(2^n)$ is the number of n-tuples of nonnegative integers $(x_1, \ldots, x_n)$ satisfying the inequality

$$2x_1 + 2^2 x_2 + \cdots + 2^n x_n \leq 2^n, \tag{5}$$

because every n-tuple satisfying (5) can be uniquely completed with an x_0-term to a solution $(x_0, x_1, \ldots, x_n)$ of the preceding equation.

If any one of the n summands on the left side of (5) is equal to 2^n, then all the other ones must be equal to zero. This yields n "trivial" solutions

$(x_1, \ldots, x_n)$:

$$(2^{n-1}, 0, \ldots, 0), \quad (0, 2^{n-2}, 0, \ldots, 0),$$
$$\ldots, \quad (0, \ldots, 0, 2, 0), \quad (0, \ldots, 0, 1).$$

Let $(x_1, \ldots, x_n)$ be any solution of (5) different from those "trivial" ones. Each term $2^i x_i$ is smaller than 2^n, which means that each x_i is an element of the set $\{0, 1, \ldots, 2^{n-i} - 1\}$. So we have 2^{n-i} independent choices for each x_i.

The number of all n-tuples $(x_1, \ldots, x_n)$ satisfying these constraints equals $2^{n-1} \cdot 2^{n-2} \cdots 2^0 = 2^{n(n-1)/2}$. Not all of them fulfill condition (5); but, in any case, all solutions of (5), except the "trivial" ones, are among them. Therefore

$$f(2^n) \leq n + 2^{n(n-1)/2}.$$

If $n \geq 3$, then the last expression has value less than $2^{n^2/2}$ (an easy exercise). So we have the upper estimate.

For the lower estimate, consider those n-tuples $(x_1, \ldots, x_n)$ in which every x_i satisfies $2^i x_i \leq 2^n/n$; for each i there exist $1 + \lfloor 2^{n-i}/n \rfloor$ such x_is. All tuples of this kind fulfill condition (5). (These are, by far, not all the solutions of (5); the total number of solutions is still greater.) Anyway,

$$f(2^n) \geq \prod_{i=1}^{n} (1 + \lfloor 2^{n-i}/n \rfloor) > \prod_{i=1}^{n} (2^{n-i}/n) = n^{-n} 2^{n(n-1)/2}. \qquad (6)$$

This number is greater than $2^{n^2/4}$ for $n \geq 19$.

To examine smaller values of n we can apply the following refinement of the above argument. For a nonnegative integer c, denote by $g(c)$ the number of pairs (x, y) of nonnegative integers such that $x + 2y \leq c$. It is not hard to check that $g(c) = \lfloor (c+2)^2/4 \rfloor$. Choose positive integers n and k with $n \geq 2k$ and consider all n-tuples $(x_1, \ldots, x_n)$ such that

$$2^{2i-1} x_{2i-1} + 2^{2i} x_{2i} \leq 2^n/k \quad \text{for } i = 1, \ldots, k; \qquad x_j = 0 \quad \text{for } j > 2k;$$

note that every such tuple satisfies condition (5). Rewriting this inequality as $x_{2i-1} + 2x_{2i} \leq \lfloor 2^{n-2i+1}/k \rfloor$ we see that for each i there exist exactly $g\left(\lfloor 2^{n-2i+1}/k \rfloor\right)$ such pairs (x_{2i-1}, x_{2i}). Thus

$$f(2^n) \geq \prod_{i=1}^{k} g\left(\lfloor 2^{n-2i+1}/k \rfloor\right).$$

To complete the proof, it suffices to show that for any n one can find an integer $k \le n/2$ so that the last product exceeds $2^{n^2/4}$: simple calculations show that e.g., $k = 2$ does the job for $n = 4, \dots, 8$ and $k = 4$ does the job for $n = 9, \dots, 18$—and this is just what we need.

Remark. Inequality (6) is true for *every* integer $n \ge 1$; as we have seen, it implies the *claimed* lower estimate for $n \ge 19$ only, so that smaller integers n require separate treatment. Yet, asymptotically, inequality (6) is significantly better than the claimed one. Note that it can be written in the form

$$f(2^n) > 2^w \quad \text{where } w = \tfrac{1}{2}n^2 - \tfrac{1}{2}n - n \cdot \log_2 n.$$

Thus if α is any number less than $1/2$, then the inequality $f(2^n) > 2^{\alpha n^2}$ holds for n large enough. The parameter $\alpha = 1/4$ from the problem statement is rather awkward; it has no clear connection with the properties of the sequence $f(2^n)$.

The problem admits several other approaches, using more advanced methods and pertaining to various domains of mathematics: combinatorics, number theory, algebra, calculus, and even geometry (lattice points in n-dimensional space); nine solutions are presented in the article [20]. Many of these methods lead to a lower estimate comparable with (6) (hence better than the proposed one), but causing cumbersome troubles with small values of n. From this point of view, the problem might have been better if it had been posed in some "asymptotic" form.

Thirty-ninth International Olympiad, 1998

1998/1

First Solution. Let Q be the point of intersection of the diagonals AC and BD.

Suppose $ABCD$ is a cyclic quadrilateral; then P is its circumcenter, and hence $\angle APB = 2 \cdot \angle QCB$, $\angle CPD = 2 \cdot \angle CBQ$. Triangles ABP and CDP have areas $\frac{1}{2}PA \cdot PB \cdot \sin \angle APB$ and $\frac{1}{2}PC \cdot PD \cdot \sin \angle CPD$. These numbers are equal because PA, PB, PC, PD are the radii of the cirumcircle and $\angle APB + \angle CPD = 2 \cdot (\angle QCB + \angle CBQ) = 180°$.

We pass to the proof of the opposite implication. Suppose that the quadrilateral $ABCD$ is not cyclic; we wish to show that triangles ABP and CDP have different areas. Of course, $PA = PB$ and $PC = PD$. By our assumption, P lies at different distances from A and C; we may assume

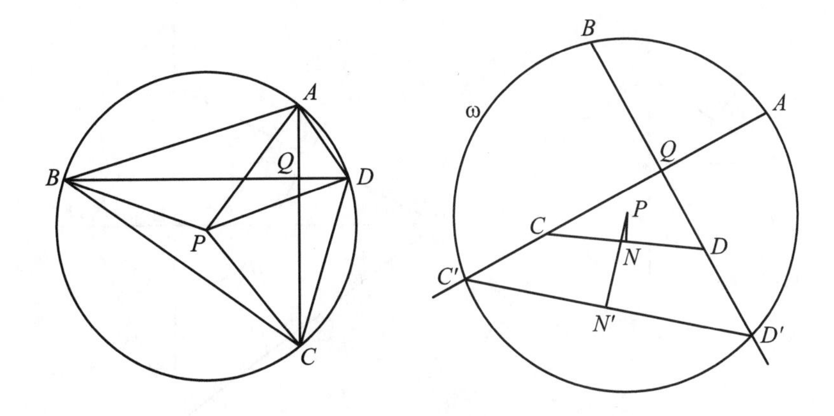

that $PA > PC$. Let ω be the circle centered at P and passing through A and B. Points C and D lie inside ω.

Let the rays QC and QD intersect ω at C' and D'. Denote the midpoints of CD and $C'D'$ by N and N'. Note that C', D' and N' lie on the other side of line CD than P and Q. Since N is the foot of the perpendicular from P to CD, we have $PN' > PN$. Lines QC and QD are perpendicular, and so $C'D' > CD$. Therefore the product $C'D' \cdot PN'$ exceeds $CD \cdot PN$. These products express the doubled areas of triangles $C'D'P$ and CDP.

Quadrilateral $ABC'D'$ is inscribed into circle ω and has perpendicular diagonals. According to what has been already shown, triangles ABP and $C'D'P$ have equal areas. So the area of triangle ABP is greater than the area of triangle CDP. The claim follows.

Second Solution. As in the first solution, let AC and BD meet at Q. Denote the midpoints of AB and CD by M and N, respectively. Without loss of generality assume $QA \leq QB$. If Q were situated closer to C than to D, then N would be the only common point of the pentagon $BMQNC$ and the perpendicular bisector of CD. The perpendicular bisector of AB meets the pentagon $AMQND$ at the single point M (unless $QA = QB$, in which case it meets that pentagon along the segment QM). Anyhow, these two bisector lines could not intersect inside the quadrilateral $ABCD$, contrary to hypothesis.

Thus the inequality $QA \leq QB$ forces $QC \geq QD$. We introduce further notation: $\varphi = \angle MBQ = \angle BQM$ and $\psi = \angle CQN = \angle NCQ$; then $\angle AMQ = 2\varphi$, $\angle QND = 2\psi$. The inequalities $QA \leq QB$ and $QC \geq QD$ imply $\varphi \leq 45°$, $\psi \leq 45°$; $QMBCN$ is a convex pentagon containing P (the point of intersection of the perpendicular bisectors of AB and CD).

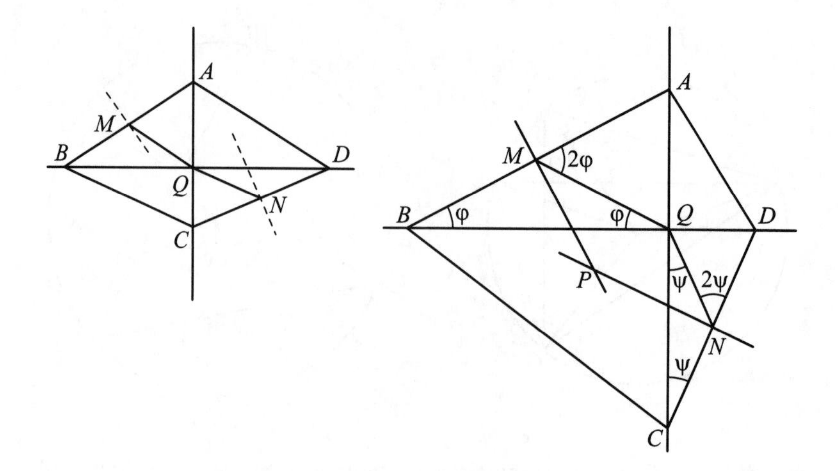

Look at the convex quadrilateral $QMPN$ (the point P can lie in any one of the sectors MQB, BQC, CQN). Since $\angle QMP = 90° - 2\varphi$, $\angle PNQ = 90° - 2\psi$, $\angle NQM = 90° + \varphi + \psi$ and the angles of $QMPN$ add up to 360°, the remaining angle has size

$$\angle MPN = 90° + \varphi + \psi.$$

Thus $\angle MPN = \angle NQM$. Note that all these equalities are a consequence of the assumptions given in the first two sentences of the problem statement and do not depend on any one of the two conditions which have to be shown equivalent.

We now write a chain of conditions, each of which is equivalent to the preceding one (the equivalences (2) $\Longleftrightarrow$ (3) and (4) $\Longleftrightarrow$ (5) follow from $\angle MPN = \angle NQM$):

$$ABCD \text{ is a cyclic quadrilateral;} \tag{1}$$

$$\varphi = \psi; \tag{2}$$

$$QMPN \text{ is a parallelogram;} \tag{3}$$

$$\text{triangle } MPN \text{ is directly similar to triangle } NQM; \tag{4}$$

$$PM/PN = QN/QM; \tag{5}$$

$$PM \cdot QM = PN \cdot QN; \tag{6}$$

$$PM \cdot AB = PN \cdot CD; \tag{7}$$

$$\text{triangles } ABP \text{ and } CDP \text{ have equal areas.} \tag{8}$$

The equivalence (1) $\Longleftrightarrow$ (8) is precisely the assertion of the problem.

1998/2

First Solution. Label the contestants 1 through a. Suppose that the jth contestant received x_j "pass" verdicts and $y_j = b - x_j$ "fail" verdicts; so there are $\binom{x_j}{2} + \binom{y_j}{2}$ pairs of examiners whose ratings coincide on the jth contestant. The total number of coincidences, T, is equal to

$$T = \sum_{j=1}^{a}\left(\binom{x_j}{2} + \binom{y_j}{2}\right) = \frac{1}{2}\sum_{j=1}^{a}(x_j^2 + y_j^2 - x_j - y_j). \tag{1}$$

On the other hand, there are, in all, $\binom{b}{2}$ pairs of examiners, the ratings of each pair coinciding for at most k contestants. Hence the total number of coincidences satisfies the estimate

$$T \le k\binom{b}{2}. \tag{2}$$

For each j we have $x_j^2 + y_j^2 \ge (x_j + y_j)^2/2 = b^2/2$; and since b is odd, this forces $x_j^2 + y_j^2 \ge (b^2 + 1)/2$. So we get from (1) and (2)

$$kb(b-1) \ge \sum_{j=1}^{a}(x_j^2 + y_j^2 - x_j - y_j) \ge a\left(\frac{b^2+1}{2} - b\right) = a \cdot \frac{(b-1)^2}{2}.$$

Cancelling $(b - 1)$ we get the required result:

$$2kb \ge a(b-1). \tag{3}$$

Second Solution. Now we also label the examiners from 1 to b. For $j = 1, 2, \ldots, a$ and $\ell = 1, 2, \ldots, b$ let $c_{\ell,j}$ be $+1$ or -1, according as the ℓth examiner rates the jth contestant as "pass" or "fail."

Let $\mathbf{v}_\ell = (c_{\ell,1}, c_{\ell,2}, \ldots, c_{\ell,a})$ be the ℓth examiner's ratings vector. Consider the dot product (see *Glossary*) of vectors $\mathbf{v}_\ell$ and $\mathbf{v}_m$ (ratings vectors of two distinct examiners):

$$\mathbf{v}_\ell \cdot \mathbf{v}_m = c_{\ell,1}c_{m,1} + c_{\ell,2}c_{m,2} + \cdots + c_{\ell,a}c_{m,a}.$$

Each summand is a $+1$ or -1, according as the ratings of the jth contestant by these two examiners coincide or not. By the condition of the problem, there are at most k $+1$s, hence at least $(a - k)$ -1s in this sum. Thus

$$\mathbf{v}_\ell \cdot \mathbf{v}_m \le 2k - a \quad \text{for } \ell, m = 1, 2, \ldots, b, \quad \ell \ne m.$$

Obviously,

$$\mathbf{v}_\ell \cdot \mathbf{v}_\ell = c_{\ell,1}^2 + c_{\ell,2}^2 + \cdots + c_{\ell,a}^2 = a \quad \text{for } \ell = 1, 2, \ldots, b.$$

Now look at the sum $\mathbf{w} = \mathbf{v}_1 + \mathbf{v}_2 + \cdots + \mathbf{v}_b$. The jth entry of $\mathbf{w}$, equal to $c_{1,j} + c_{2,j} + \cdots + c_{b,j}$, is an odd integer (because b is odd). Therefore $\mathbf{w}\cdot\mathbf{w}$ is the sum of the squares of a odd integers, hence a number not less than a. Using these estimates we obtain

$$a \le \mathbf{w}\cdot\mathbf{w} = \left(\sum_{\ell=1}^{b} \mathbf{v}_\ell\right)\cdot\left(\sum_{\ell=1}^{b} \mathbf{v}_\ell\right) = \sum_{\ell=1}^{b} \mathbf{v}_\ell\cdot\mathbf{v}_\ell + 2\sum_{\ell<m} \mathbf{v}_\ell\cdot\mathbf{v}_m$$
$$\le ab + 2(2k-a)\binom{b}{2}.$$

Rewriting this as $0 \le a(b-1) + (2k-a)b(b-1)$ and cancelling $(b-1)$ we get the claimed inequality (3).

1998/3

For $n = 1$ we have $d(1^2)/d(1) = 1$; thus 1 is a feasible value of k. Consider an arbitrary integer $n \ge 2$; let $n = p_1^{\alpha_1} p_2^{\alpha_2} \cdots p_r^{\alpha_r}$ be its prime factorization (with exponents $\alpha_i \ge 1$). Each positive divisor of n has the form $p_1^{\lambda_1} p_2^{\lambda_2} \cdots p_r^{\lambda_r}$ where $0 \le \lambda_i \le \alpha_i$; so we have $\alpha_i + 1$ possible choices of λ_i, and hence $d(n) = (\alpha_1+1)(\alpha_2+1)\cdots(\alpha_r+1)$. Since $n^2 = p_1^{2\alpha_1} p_2^{2\alpha_2} \cdots p_r^{2\alpha_r}$, we obtain

$$\frac{d(n^2)}{d(n)} = \frac{2\alpha_1+1}{\alpha_1+1}\cdot\frac{2\alpha_2+1}{\alpha_2+1}\cdots\frac{2\alpha_r+1}{\alpha_r+1}. \qquad (*)$$

(This formula is also valid for $n = 1$, with a "void" product on the right side, evaluating to 1.) The problem reduces to the following: what positive integers k can be represented as a product of a finite number of fractions of type $(2\alpha+1)/(\alpha+1)$?

If the product on the right side of $(*)$ has an integer value, it must be odd. We will prove by induction the converse statement: every odd integer $k \ge 1$ is an available value of the product in $(*)$. This is true, in particular, for $k = 1$ (induction base).

Fix an odd integer $k \ge 3$ and assume that every positive odd integer less than k can be written as a product of fractions $(2\alpha+1)/(\alpha+1)$. The even integer $k+1$ equals $2^r\ell$ for some integers $r, \ell \ge 1$, with ℓ odd. Since $k+1 \ge 2\ell$, we see that ℓ is smaller than k; hence it is a product of fractions $(2\alpha+1)/(\alpha+1)$, by induction hypothesis. Thus it will be enough to show that also k/ℓ can be represented as a product of such fractions.

Let $k - \ell = j$. From the equation $k+1 = 2^r \ell = 2^r(k-j)$ we compute $k = (2^r j + 1)/(2^r - 1)$. Therefore

$$\frac{k}{\ell} = \frac{k}{k-j} = \left(1 - \frac{j}{k}\right)^{-1} = \left(1 - \frac{j(2^r-1)}{2^r j + 1}\right)^{-1} = \left(\frac{j+1}{2^r j + 1}\right)^{-1}$$
$$= \frac{2j+1}{j+1} \cdot \frac{4j+1}{2j+1} \cdot \frac{8j+1}{4j+1} \cdots \frac{2^r j + 1}{2^{r-1} j + 1},$$

as required. This completes the induction step.

Conclusion: the set of possible integer values of the ratio $d(n^2)/d(n)$ coincides with the set of all positive odd integers.

1998/4

Suppose a and b are positive integers such that the number $A = a^2b + a + b$ is divisible by $B = ab^2 + b + 7$. Then also the difference $Ab - Ba = b^2 - 7a$ is divisible by B.

First, consider the case where $Ab - Ba = b^2 - 7a = 0$. Then a and b are integers of the form

$$a = 7c^2, \quad b = 7c \quad \text{for some integer } c \geq 1. \tag{1}$$

It is easy to check that every such pair (a, b) satisfies the postulated condition (in fact, $A = Bc$ in this case).

Now assume $b^2 - 7a \neq 0$. This number is divisible by $B = ab^2 + b + 7$ and so $ab^2 + b + 7 \leq |b^2 - 7a|$; i.e.,

$$b^2 - 7a \geq ab^2 + b + 7 \quad \text{or} \quad 7a - b^2 \geq ab^2 + b + 7.$$

The first alternative is obviously impossible; and the second one, rewritten as $(7 - b^2)a \geq b^2 + b + 7$, shows that b must be either 1 or 2.

If $b = 1$, the condition of the problem says that $a + 8$ has to divide $a^2 + a + 1$. Thus the number

$$\frac{a^2 + a + 1}{a + 8} = a - 7 + \frac{57}{a + 8}$$

has to be an integer. In view of the prime factorization $57 = 3 \cdot 19$ we get that $a = 11$ or $a = 49$. The two pairs

$$a = 11, \quad b = 1 \quad \text{and} \quad a = 49, \quad b = 1 \tag{2}$$

satisfy the requirement. If $b = 2$, we analogously infer that

$$\frac{2a^2 + a + 2}{4a + 9} = \frac{1}{2}\left(a - \frac{1}{4}\left(7 - \frac{79}{4a + 9}\right)\right)$$

has to be an integer; and since 79 has no divisor of the form $4a+9$, this subcase yields no solution. Thus the complete solution is described by formulas (1) and (2).

1998/5

The segments BK and BM are equal tangents to the incircle of triangle ABC. Hence, by the tangent-chord theorem (and because $RS \parallel MK$),

$$\angle BMR = \angle MKL = \angle BSK \quad \text{and} \quad \angle SKB = \angle LMK = \angle MRB.$$

These angle equalities show that triangle BMR is directly similar to BSK, and consequently $BR/BM = BK/BS$.

On the segment BI (perpendicular to MK, hence also to RS) lay off $BJ = BM = BK$. The last equality now rewrites as $BR/BJ = BJ/BS$. Thus $BJ = \sqrt{BR \cdot BS}$, showing that RJS is a right triangle with hypotenuse RS.

The common hypotenuse BI of the right triangles BMI and BKI is longer than their legs BM and BK; thus $BI > BJ$. Consequently, J lies inside the triangle RIS, and therefore the angle RIS is acute.

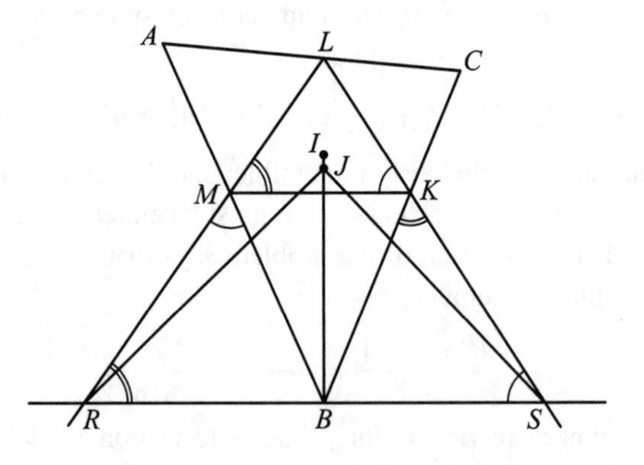

Remark. Vector calculus can be also used in the solution of this problem. To prove that the angle RIS is acute, it is enough to show that the dot product of vectors $\overrightarrow{IR}$ and $\overrightarrow{IS}$ is positive (see *Glossary*). Readers fond of calculations may like to verify that this product is equal to r^2, where r is the inradius of triangle ABC.

1998/6

Let $f(1) = a$. In the given equation

$$f(t^2 f(s)) = s(f(t))^2 \tag{1}$$

set, first, $t = 1$, and then $s = 1$, to obtain

$$f(f(s)) = a^2 s \quad \text{and} \quad f(at^2) = (f(t))^2. \tag{2}$$

Take any $x, y \in \mathrm{N}$. In the first equation of (2) set $s = ay^2$:

$$f(f(ay^2)) = a^3 y^2. \tag{3}$$

Go back to (1), with $t = x$, $s = f(ay^2)$:

$$f(x^2 f(f(ay^2))) = f(ay^2)(f(x))^2.$$

Using relation (3) and the second equation of (2) we rewrite this as

$$f(x^2 \cdot a^3 y^2) = (f(y))^2 (f(x))^2.$$

Applying once more the second equation of (2) (with $t = axy$) we see that the left side of the last equality is equal to $(f(axy))^2$. Hence

$$f(axy) = f(x)f(y). \tag{4}$$

This is true for every $x, y \in \mathrm{N}$. Taking $y = 1$ we get $f(ax) = af(x)$; thus (4) becomes

$$af(xy) = f(x)f(y). \tag{5}$$

Hence, by induction,

$$a^{k-1} f(x^k) = (f(x))^k \quad \text{for } k = 1, 2, 3, \dots. \tag{6}$$

We claim that all values of f are integers divisible by a. Fix an $x \in \mathrm{N}$ and take any prime number p. Let p^α be the highest power of p dividing a, and let p^β be the highest power of p dividing $f(x)$. Formula (6) shows that $(k - 1)\alpha \le k\beta$ for $k = 1, 2, 3, \dots$. This infinite system of inequalities is satisfied only if $\alpha \le \beta$. Since p is an arbitrary prime, we conclude that $f(x)$ is divisible by a. And since also the choice of x was arbitrary, it follows that all values of f are divisible by a, as claimed. So the formula

$$g(x) = \frac{f(x)}{a} \quad \text{for } x \in \mathrm{N}$$

defines a function $g\colon \mathrm{N} \to \mathrm{N}$. Note that $g(1) = 1$. Relation (5) yields for the function g the equation

$$g(xy) = g(x)g(y) \quad \text{for } x, y \in \mathrm{N}, \tag{7}$$

while the first equation of (2) implies $g(ag(s)) = as$; for $s = 1$ we obtain $g(a) = a$. Thus $as = g(ag(s)) = g(a)g(g(s)) = ag(g(s))$, whence

$$g(g(s)) = s \quad \text{for } s \in \mathrm{N}. \tag{8}$$

And conversely, it is not hard to check that if $g\colon \mathrm{N} \to \mathrm{N}$ is any function fulfilling conditions (7) and (8), and if $a \in \mathrm{N}$ is any integer with $g(a) = a$, then the formula $f(x) = ag(x)$ defines a function satisfying the original equation (1).

We now show that for every prime p the value $g(p)$ is also a prime. In view of (8), g is a one-to-one function; so $g(p) \neq g(1) = 1$. Assume $g(p) = uv(u, v \in \mathrm{N})$. By (7) and (8), $g(u)g(v) = g(uv) = g(g(p)) = p$; so $g(u) = 1$ or $g(v) = 1$, and hence $u = g(g(u)) = 1$ or $v = g(g(v)) = 1$. This means that $g(p)$ is indeed a prime integer.

Thus g maps the set of all primes into itself. From equation (7) we obtain $g(1998) = g(2 \cdot 3^3 \cdot 37) = qr^3s$, where $q = g(2)$, $r = g(3)$, $s = g(37)$ are three distinct primes. So the value of the product qr^3s is not less than $3 \cdot 2^3 \cdot 5 = 120$. Hence $f(1998) = a \cdot g(1998) \geq 120$.

It is reasonable to conjecture that 120 is the minimum we seek. To justify that, it will be enough to find a function f satisfying equation (1) with $f(1998) = 120$. Now, setting $g(1) = 1$, $g(2) = 3$, $g(3) = 2$, $g(5) = 37$, $g(37) = 5$ and, say, $g(p) = p$ for all primes $p \neq 2, 3, 5, 37$, and defining $g(x) = (g(p_1))^{\alpha_1} \cdots (g(p_n))^{\alpha_n}$ for $x = p_1^{\alpha_1} \cdots p_n^{\alpha_n}$, we get conditions (7) and (8) satisfied.

Taking $a = 1$ (which is equivalent to defining $f(x) = g(x)$) we obtain a function f satisfying equation (1). Moreover, by virtue of (7), $f(1998) = g(2 \cdot 3^3 \cdot 37) = 3 \cdot 2^3 \cdot 5 = 120$. This proves that, indeed, 120 is the minimum value available as $f(1998)$.

Fortieth International Olympiad, 1999

1999/1

Let S be a set satisfying the given condition and let W be the smallest convex polygon containing S. All vertices of W are points of S. First, we show that all sides of W have equal lengths. Suppose not; then there are two adjacent sides AB, BC of different lengths. Let ℓ be the perpendicular bisector of AC; by hypothesis, ℓ is an axis of symmetry of S. Let $B' \in S$ be the mirror image of B across ℓ; note that B and B' do not coincide. So $AB'BC$ (or $ABB'C$) is a nondegenerated trapezoid with vertex B' outside W—a contradiction with the definition of W. Thus, indeed, W is a polygon with equal sides.

Now we prove that, in fact, W is a regular polygon. When it is a triangle, there is nothing more to show. Thus assume W has at least four vertices. It has to be shown that all internal angles are equal. Again, suppose this is not the case; then there are two consecutive angles ABC, BCD of different sizes; let e.g., $\angle ABC > \angle BCD$. The perpendicular bisector of BC is an axis of symmetry of S; the point $A' \in S$, symmetric to A with respect to that line, lies outside W—a contradiction again. So W is a regular polygon.

The last claim is that there are no points in S other than the vertices of W. Suppose P is such a point; it lies inside the circumcircle of W. Let A be any vertex of W and let m be the perpendicular bisector of AP, hence a symmetry axis of S. It does not pass though the circumcenter of W, so it divides the circle into two unequal arcs. Point A lies on the minor arc; and there is at least one vertex E of W on the major arc. Its symmetric image across m (a point of S) lies outside the circumcircle, hence outside W, yielding a contradiction once more.

Thus we have shown that S is precisely the set of all vertices of a regular polygon. And conversely, it is clear that the set of vertices of any regular polygon satisfies the condition from the problem statement.

1999/2

First Solution. Let $x_1, \ldots, x_n$ be any nonnegative numbers. Denote their sum by S. Then

$$S^2 = A + B$$

where

$$A = \sum_{1 \le i \le n} x_i^2$$

$$B = 2 \sum_{1 \le i < j \le n} x_i x_j.$$

Obviously $x_i^2 + x_j^2 \leq A$ for $i < j$. Therefore

$$\begin{aligned}\sum_{1\leq i<j\leq n} x_i x_j (x_i^2 + x_j^2) &\leq \sum_{1\leq i<j\leq n} x_i x_j \cdot A \\ &= \frac{AB}{2} = \frac{(A+B)^2 - (A-B)^2}{8} \\ &\leq \frac{(A+B)^2}{8} = \frac{S^4}{8}. \qquad (1)\end{aligned}$$

This shows that the inequality is satisfied with the constant $C = 1/8$.

In the first line of (1), equality holds if, for each pair of distinct indices i, j, either $x_i x_j = 0$ (i.e., one of the numbers x_i, x_j is zero) or $x_i^2 + x_j^2 = A$. In the latter case all the remaining numbers in $(x_1, \ldots, x_n)$ must be zeros. Thus we have equality in the first line of (1) if and only if at most two numbers in $(x_1, \ldots, x_n)$ are different from zero.

Consider an n-tuple with this property, i.e., consisting of two nonnegative numbers a, b and $n-2$ zeros. Then $A = a^2 + b^2$, $B = 2ab$. Equality in the last line of (1) requires that $A = B$; and this is the case if and only if $a = b$.

The answer follows: the minimum constant C is $1/8$. For $C = 1/8$ the inequality turns into equality if and only if $(x_1, \ldots, x_n)$ is a permutation of $(a, a, 0, \ldots, 0)$, with an arbitrary $a \geq 0$.

Second Solution. When all the x_is are zero, the inequality is fulfilled. For the sequel we restrict attention to n-tuples $(x_1, \ldots, x_n)$ of nonnegative numbers with a positive sum. The expressions on both sides of the inequality are homogeneous polynomials of degree 4. Thus if a constant C is "good" for every nonnegative number $x_1, \ldots, x_n$ summing to 1, then it is "good" for every n nonnegative number with an arbitrary positive sum. This reduces the problem to the following: determine the least constant C such that the inequality

$$F(x_1, \ldots, x_n) = \sum_{1\leq i<j\leq n} x_i x_j (x_i^2 + x_j^2) \leq C$$

is satisfied for any nonnegative numbers $x_1, \ldots, x_n$ whose sum equals 1.

The formula defining F can be rewritten as

$$F(x_1, \ldots, x_n) = \sum_{i<j} x_i^3 x_j + \sum_{i>j} x_i^3 x_j = \sum_{i=1}^{n} x_i^3 (1 - x_i). \qquad (2)$$

The function $f(x) = x^2 - x^3$ is increasing on $[0, 1/2]$; so

$$\text{if} \quad 0 \le x \le y \le 1/2, \quad \text{then } x^2 - x^3 \le y^2 - y^3. \tag{3}$$

If all the x_is are numbers from the interval $[0, 1/2]$, we apply (3) to each of them, taking $y = 1/2$:

$$F(x_1, \dots, x_n) = \sum_{i=1}^{n} x_i(x_i^2 - x_i^3) \le \sum_{i=1}^{n} x_i \left(\frac{1}{4} - \frac{1}{8}\right) = \frac{1}{8}.$$

Equality holds in this estimate if and only if each nonzero x_i satisfies the equation $x_i^2 - x_i^3 = 1/8$; in the interval $[0, 1/2]$ this equation is fulfilled only for $x_i = 1/2$. And since $\sum x_i = 1$, equality holds if and only if two of the x_is are equal to $1/2$ and all the others are 0.

We are left with the case where one of the x_is is greater than $1/2$; say, $x_1 > 1/2$. Now we use property (3) with $x = x_i$ $(i = 2, \dots, n)$ and $y = x_2 + \dots + x_n = 1 - x_1 < 1/2$:

$$\begin{aligned} F(x_1, \dots, x_n) &= x_1^3(1 - x_1) + \sum_{i=2}^{n} x_i^3(1 - x_i) = x_1^3 y + \sum_{i=2}^{n} x_i(x_i^2 - x_i^3) \\ &\le x_1^3 y + \sum_{i=2}^{n} x_i(y^2 - y^3) = x_1^3 y + y(y^2 - y^3) \\ &= x_1 y(x_1^2 + y^2). \end{aligned}$$

The last value is smaller than $1/8$ when $y < 1/2$ and $x_1 > 1/2$ (easy verification).

In both cases we have obtained the estimate $F(x_1, \dots, x_n) \le 1/8$, becoming an equality only for $(x_1, \dots, x_n)$ with $1/2$ on two positions and otherwise 0. Thus $C = 1/8$ is the least "good" constant. Dismissing the constraint $\sum x_i = 1$ we obtain, in view of homogeneity, the characterization of equality instances (for $C = 1/8$) as in the first solution.

Third Solution. Again we reduce the problem to considering vectors $(x_1, \dots, x_n)$ with nonnegative entries summing to 1. Take any such vector. Suppose it has at least three positive terms; we may assume, relabeling if necessary, $x_1 \ge x_2 \ge \dots \ge x_k > 0$, $x_{k+1} = \dots = x_n = 0$ $(k \ge 3)$. Write $w = x_{k-1} + x_k$; so $w \le 2/3$.

Denote the value $F(x_1, \dots, x_n)$ by W. We modify the vector by replacing the last two positive terms x_{k-1} and x_k respectively by w and 0. Let

W' be the value of F on this new vector:

$$F(x_1, \dots, x_{k-2}, w, 0, \underbrace{0, \dots, 0}_{n-k}) = W'.$$

It can be checked (this is a matter of a simple calculation, using (2)) that

$$W' - W = 3x_{k-1}x_k w(1-w) - x_{k-1}x_k(w^2 - 2x_{k-1}x_k).$$

The last expression has value greater than

$$3x_{k-1}x_k(w - w^2) - x_{k-1}x_k w^2 = x_{k-1}x_k w(3 - 4w) > 0.$$

The new vector is also built of numbers whose sum is 1. Rearrange its terms decreasingly. If it has at least three nonzero terms, we repeat the reasoning. Starting from an arbitrary vector $(x_1, \dots, x_n)$ $(x_i \geq 0, \sum x_i = 1)$ with at least three positive terms, we successively reduce their number, raising the value of F in each step. Eventually we obtain a vector of the form $(a, b, 0, \dots, 0)$; $a, b \geq 0$, $a + b = 1$; and it is no problem to verify that $F(a, b, 0, \dots, 0) \leq 1/8$, equality holding only for $a = b = 1/2$. The conclusion is as in the two previous solutions.

1999/3

Let $ABCD$ be the board. Color its squares black and white like a chessboard, the diagonal AC consisting of white squares, and BD of black ones. Let $f(n)$ be the smallest possible value of N, as in the problem statement. Define $f_w(n)$ as the minimum number of white squares that must be *marked* in order that every black square on the board be adjacent to at least one *marked* white square. Define $f_b(n)$ analogously, with the roles of "black" and "white" interchanged. These roles are fully symmetric because n is even. Thus $f_w(n) = f_b(n)$; obviously, $f_w(n) + f_b(n) = f(n)$.

Consider those "white bishop" lines (i.e., oblique lines of white squares) that run in parallel to BD; label them consecutively $\ell_1, \ell_2, \ell_3, \ell_4, \dots,$ so that ℓ_1 reduces to the single white square having A as a vertex. On each line ℓ_i with an odd index i, put a cross on every second square, starting and ending at squares adjacent to edges of the board.

Let $n = 2k$. Odd-indexed lines ℓ_i running between A and the diagonal BD have lengths 1, 5, 9, etc. (1 mod 4); on these lines we have put $1 + 3 + 5 + \cdots + h$ crosses, where h is the greatest odd integer not exceeding k.

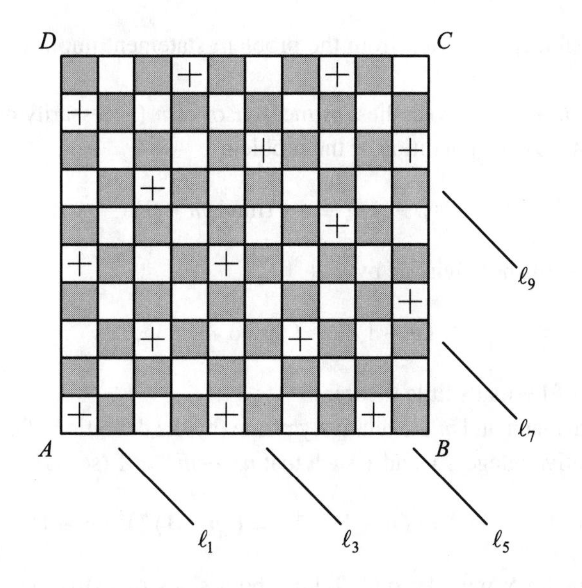

Odd-indexed lines ℓ_i running between C and the diagonal BD have lengths 3, 7, 11, etc. (3 mod 4); on these, we have put $2 + 4 + 6 + \cdots + h'$ crosses, where h' is the greatest even integer not exceeding k.

Let $\mathcal{S}$ be the set of all squares with crosses. The total number of squares in $\mathcal{S}$ equals $k(k+1)/2$ (the sum of all positive integers, even or odd, not exceeding k). This set plays a double role in the solution of the problem.

First, note that there is no black square adjacent to two (or more) squares from $\mathcal{S}$. Thus the number of black squares that have to be *marked* (to fulfill the condition of the problem) cannot be smaller than the number of squares in $\mathcal{S}$. Hence $f_b(n) \geq k(k+1)/2$.

Second, it is not hard to see that every black square is adjacent to some square from $\mathcal{S}$. Thus, if we *mark* all squares in $\mathcal{S}$, this will do the job required from the *marking* of white squares. Therefore $f_w(n) \leq k(k+1)/2$.

Hence we obtain the outcome: $f_w(n) = f_b(n) = k(k+1)/2$, so that $f(n) = f_w(n) + f_b(n) = k(k+1)$, where $k = n/2$. This is the minimum value of N.

1999/4

Let (n, p) be one of the pairs sought. If $n = 1$, then of course every p is good. Let now $n > 1$. If $p = 2$, we get that n has to be a divisor of 2; i.e., $n = 2$. For the sequel assume that p is an odd prime and that $n > 1$.

The divisibility condition from the problem statement implies that n must be odd.

Let $m+1$ be the smallest prime divisor of n (necessarily odd, so that m is even). By the condition of the problem,

$$(p-1)^n \equiv -1 \pmod{m+1}. \tag{1}$$

Since $p-1$ is not divisible by $m+1$,

$$(p-1)^m \equiv 1 \pmod{m+1}, \tag{2}$$

in virtue of Fermat's little theorem.

Note that n and m are relatively prime (by the definition of m). So there exist positive integers x and y such that $nx - my = 1$ (see *Glossary*). Thus

$$(p-1)^{nx} = (p-1)^{my+1} = \left((p-1)^m\right)^y (p-1); \tag{3}$$

m is even, so x must be odd. Taking both sides of (3) (mod $m+1$) we obtain, in view of (1) and (2), $-1 \equiv p-1 \pmod{m+1}$; equivalently, $p \equiv 0 \pmod{m+1}$. And this means that the primes $m+1$ and p are equal; thus n is divisible by p.

Until this point we did not need the condition $n \le 2p$ (see the Remark). Using this assumption and keeping in mind that n is odd we get $n = p$. The divisibility condition of the problem now says that p^{p-1} has to be a divisor of the number

$$(p-1)^p + 1 = \left(\sum_{j=0}^{p-2} \binom{p}{j} p^{p-j}(-1)^j + p^2 - 1\right) + 1.$$

Note that all the terms $\binom{p}{j} p^{p-j}(-1)^j$ for $j = 0, \ldots, p-2$ are divisible by p^3 (because the binomial coefficients are divisible by p for $j = 1, \ldots, p-1$). Thus the whole expression represents an integer divisible by p^2, but not by p^3. Consequently p^{p-1} divides p^2. Since p is odd, we obtain $n = p = 3$ as the last pair we sought.

Recalling the outcomes from the first paragraph, we conclude that the complete solution is constituted by the following pairs (n, p): $(3, 3)$, $(2, 2)$ and $(1, p)$ with p being an arbitrary prime.

Remark. With considerably more effort, one can solve this problem without the premise $n \le 2p$, which is therefore redundant in the problem statement; the solution pairs (n, p) will be the same. For this purpose, one can prove by induction on $k = 0, 1, 2, \dots$ that, for any odd prime p,

$$(p-1)^{p^k} + 1 \quad \text{is divisible by } p^{k+1}, \text{ but not by } p^{k+2}. \tag{4}$$

This is obvious for $k = 0$. Assuming $(p-1)^{p^k} = ap^{k+1} - 1$ for a certain integer $k \geq 0$, where a is an integer nondivisible by p, we have $(p-1)^{p^{k+1}} + 1 = \left(ap^{k+1} - 1\right)^p + 1$. Expanding the power binomially (and knowing that the binomial coefficients $\binom{p}{1}, \dots, \binom{p}{p-1}$ are divisible by p) we obtain without much difficulty the induction claim.

Going back to the problem, in the case where p is an odd prime and $n > 1$, we arrive at the conclusion that p divides n, as in the solution above (without assuming any bound on n). Let $n = p^k s$ with s nondivisible by p. Then

$$(p-1)^n + 1 = \left((p-1)^{p^k}\right)^s + 1 = \left((p-1)^{p^k} + 1\right) \cdot S, \tag{5}$$

where

$$S = \sum_{j=0}^{s-1} \left(-(p-1)^{p^k}\right)^j \equiv \sum_{j=0}^{s-1} (-(-1))^j \equiv s \pmod{p}. \tag{6}$$

It follows from (4), (5), and (6) that the number $(p-1)^n + 1$ is not divisible by p^{k+2}; and by the condition of the problem, this number is divisible by $p^{k(p-1)}$. Hence $k(p-1) \leq k+1$, implying $p \leq 2 + (1/k)$, so that $p = 3$ and $k = 1$, i.e., $n = 3s$.

The condition of the problem becomes: n^2 is a divisor of $2^n + 1$; and this is exactly the contents of problem 1990/3. But now we already know that $n = 3s$ with s coprime to 3. Thus we are in the position of having reached formula (3) from the solution of 1990/3; so it remains to reproduce the final passage of that solution to conclude $s = 1$ and, eventually, $n = 3$.

1999/5

First Solution. Let AN meet Γ_2 again at E and let Γ_1 intersect Γ_2 at X and Y. Then $AC \cdot AM = AX \cdot AY = AE \cdot AN$, i.e., $AC/AE = AN/AM$; so the triangles AEC and AMN (with a common angle A) are similar. Hence $\angle AEC = \angle AMN$.

Draw the ray AP tangent to Γ at A, with P lying on the other side of line AB than M; angles PAN and AMN are equal by the tangent-chord theorem. In view of the previous equality, this implies $\angle PAN = \angle AEC$, and consequently lines AP and CE are parallel.

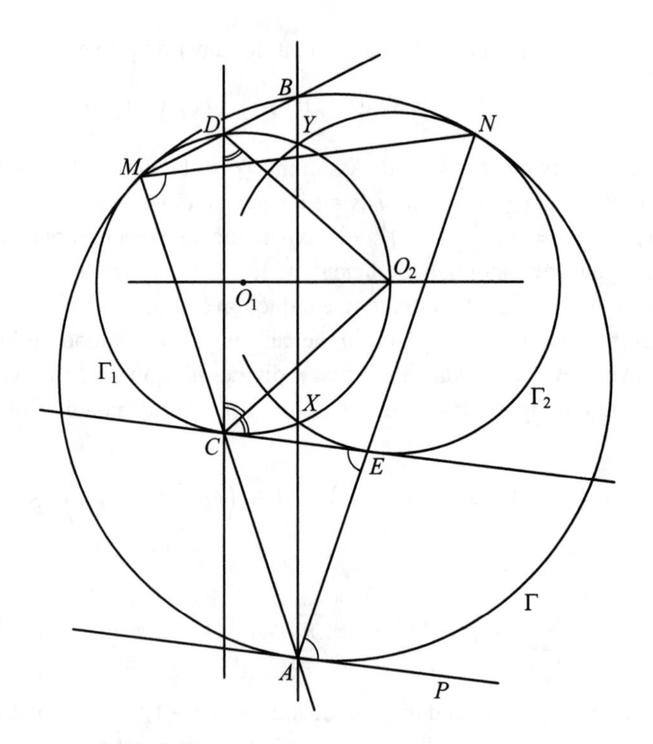

Consider the homothety centered at M and mapping circle Γ onto Γ_1. Line AP is mapped onto CE. Since AP is tangent to Γ, we infer that CE is tangent to Γ_1. Analogously, the homothety centered at N and mapping Γ onto Γ_2 takes line AP also onto CE, which is therefore tangent to Γ_2. It follows that CE is a common tangent of both small circles (Γ_1 and Γ_2).

Chord CD of Γ_1 (the image of AB in the first homothety) is parallel to AB, hence perpendicular to the line through O_1 and O_2, the centers of Γ_1 and Γ_2. Thus O_2CD is an isosceles triangle: $O_2C = O_2D$, and therefore $\angle O_2CD = \angle CDO_2 = \angle ECO_2$; the latter equality follows again from the tangent-chord theorem for circle Γ_1. Hence CO_2 bisects angle ECD. And since CE is tangent to Γ_2, we see that also CD is tangent to Γ_2, as asserted (note that the proof is fully case-independent).

Second Solution. Denote the centers of $\Gamma, \Gamma_1, \Gamma_2$ by O, O_1, O_2, and their radii by r, r_1, r_2. Let X be one of the points of intersection of Γ_1 and Γ_2. Let lines AB and CD cut O_1O_2 at K and L. Finally, define U and W as the feet of perpendiculars from O to lines O_1O_2 and AB, respectively. We proceed by vector calculus.

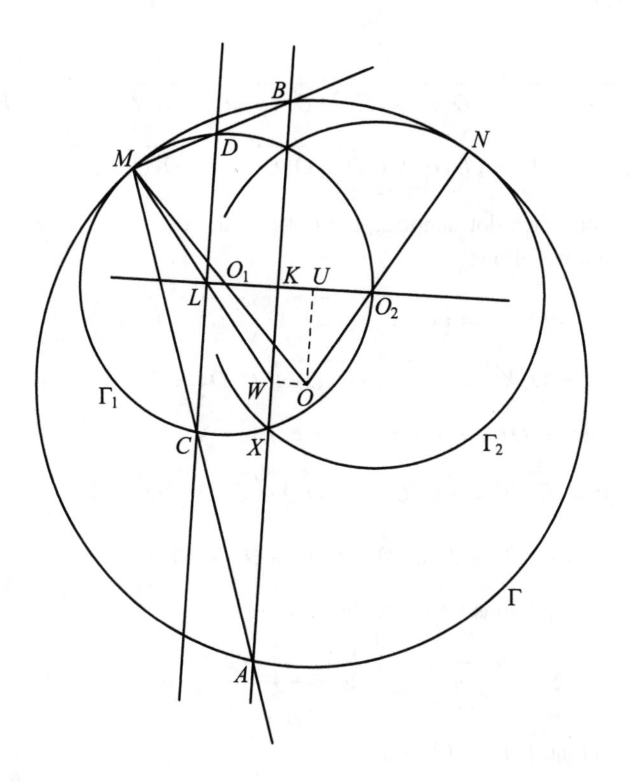

As in the first solution, we consider the homothety in ratio r_1/r, centered at M and mapping circle Γ onto Γ_1. It maps AB onto CD; hence $CD \perp O_1O_2$. The center O of Γ goes to O_1. The image of line OW is the line parallel to OW and passing through O_1; i.e., line O_1O_2. Thus W (the intersection of lines AB and OW) is sent to L (the intersection of CD and O_1O_2), and we get

$$\overrightarrow{O_1L} = \frac{r_1}{r} \cdot \overrightarrow{OW} = \frac{r_1}{r} \cdot \overrightarrow{UK} = \frac{r_1}{r} \cdot \left(\overrightarrow{O_1K} - \overrightarrow{O_1U}\right). \tag{1}$$

Points O_1, O_2, K, L, U can be situated on line O_1O_2 in various orders; but in any case,

$$\overrightarrow{O_1K} = k \cdot \overrightarrow{O_1O_2}, \qquad \overrightarrow{O_1U} = u \cdot \overrightarrow{O_1O_2} \tag{2}$$

for some real coefficients u and k. We are now going to express their values in terms of the radii of the given circles.

Vectors $\overrightarrow{KX}$ and $\overrightarrow{UO}$ are perpendicular to vector $\overrightarrow{O_1O_2}$, so that

$$\overrightarrow{KX} \cdot \overrightarrow{O_1O_2} = 0, \quad \overrightarrow{UO} \cdot \overrightarrow{O_1O_2} = 0,$$

and

$$\overrightarrow{O_1X}\cdot\overrightarrow{O_1O_2} = \left(\overrightarrow{O_1K}+\overrightarrow{KX}\right)\cdot\overrightarrow{O_1O_2} = k\cdot\overrightarrow{O_1O_2}\cdot\overrightarrow{O_1O_2} = kr_1^2,$$
$$\overrightarrow{O_1O}\cdot\overrightarrow{O_1O_2} = \left(\overrightarrow{O_1U}+\overrightarrow{UO}\right)\cdot\overrightarrow{O_1O_2} = u\cdot\overrightarrow{O_1O_2}\cdot\overrightarrow{O_1O_2} = ur_1^2,$$

with the boldface dot denoting the dot product of vectors (see *Glossary*). The equations follow:

$$r_2^2 = O_2X^2 = \left(\overrightarrow{O_1X}-\overrightarrow{O_1O_2}\right)\cdot\left(\overrightarrow{O_1X}-\overrightarrow{O_1O_2}\right)$$
$$= O_1X^2 - 2\cdot\overrightarrow{O_1X}\cdot\overrightarrow{O_1O_2} + O_1O_2^2 = r_1^2 - 2kr_1^2 + r_1^2$$

and (in view of $OO_2 = r - r_2$, $OO_1 = r - r_1$):

$$(r-r_2)^2 = O_2O^2 = \left(\overrightarrow{O_1O}-\overrightarrow{O_1O_2}\right)\cdot\left(\overrightarrow{O_1O}-\overrightarrow{O_1O_2}\right)$$
$$= O_1O^2 - 2\cdot\overrightarrow{O_1O}\cdot\overrightarrow{O_1O_2} + O_1O_2^2 = (r-r_1)^2 - 2ur_1^2 + r_1^2.$$

From these equations we compute

$$k = \frac{2r_1^2 - r_2^2}{2r_1^2}, \qquad u = \frac{2r_1^2 - 2rr_1 + 2rr_2 - r_2^2}{2r_1^2}. \tag{3}$$

From formulas (1) and (2) we obtain

$$\overrightarrow{O_2L} = \overrightarrow{O_1L} - \overrightarrow{O_1O_2} = \frac{r_1}{r}\cdot\left(\overrightarrow{O_1K}-\overrightarrow{O_1U}\right) - \overrightarrow{O_1O_2}$$
$$= \left[\frac{r_1}{r}\cdot(k-u) - 1\right]\cdot\overrightarrow{O_1O_2}.$$

On substituting the values from (3), a short calculation shows that the number in square brackets equals $-r_2/r_1$. Thus $O_2L = (r_2/r_1)\cdot O_1O_2 = r_2$. This means that L is a point of circle Γ_2, and consequently line CD is tangent to this circle.

1999/6

Suppose a function $f\colon \mathrm{R}\to\mathrm{R}$ satisfies the equation. Denote $f(0)$ by c. Setting in the equation $x = y = 0$ we get $f(-c) = f(c) + c - 1$, so that $c \neq 0$.

Let W be the set of all values taken by f. In the equation, write w in place of $f(y)$ (and z in place of x); the equation becomes

$$f(z-w) = f(w) + zw + f(z) - 1 \quad \text{for } z\in\mathrm{R},\ w\in W. \tag{1}$$

Setting $z = w \in W$ we obtain $c = 2f(w) + w^2 - 1$, whence

$$f(w) = \frac{c + 1 - w^2}{2} \quad \text{for } w \in W. \tag{2}$$

We will prove that the same formula defines f for all values of the variable $x \in \mathrm{R}$. For this purpose we first show that every real number can be written as the difference of two numbers from W; this is the key step in the solution.

For $w = c$ equation (1) takes the form

$$f(z - c) - f(z) = cz + f(c) - 1 \quad \text{for } z \in \mathrm{R}. \tag{3}$$

Fix an arbitrary $x \in \mathrm{R}$ and set $z = (x + 1 - f(c))/c$. For this value of z the right side of (3) is equal to x. Denoting by u and v the terms on the left side of (3) we have the equality $x = u - v$; clearly, u and v are elements of W.

Using equation (1) and then formula (2) (for $w = u$ and for $w = v$) we obtain

$$\begin{aligned} f(x) &= f(u - v) = f(v) + uv + f(u) - 1 \\ &= \frac{c + 1 - u^2}{2} + uv + \frac{c + 1 - v^2}{2} - 1 \\ &= c - \frac{(u - v)^2}{2} = c - \frac{x^2}{2}. \end{aligned} \tag{4}$$

This holds for every $x \in \mathrm{R}$. Comparing formulas (2) and (4) we conclude that $c = (c + 1)/2$, i.e., $c = 1$. Thus, finally,

$$f(x) = 1 - \frac{x^2}{2} \quad \text{for all } x \in \mathrm{R};$$

and it is easy to verify that this function indeed satisfies the given equation.

Results

The following tables give a summary of results of the IMOs from 1986 to 1999, listed by countries. At all these IMOs, each participating country's team normally consisted of six contestants. In those cases where the team size was less than six, this is indicated by the number in parentheses.

Each contestant's solution to each problem was allocated a score out of a maximum of seven points. Thus the joint score of all contestants from one country could reach a maximum of 252 points.

According to the IMO Regulations, approximately a half of the total number of contestants at each IMO received prizes; specifically, First, Second and Third, in the approximate ratio 1:2:3. Starting from the 1988 IMO, prizes have been called Medals (Gold, Silver, Bronze); moreover, each contestant who received no medal and had solved at least one problem with the top mark of seven points, received an Honorable Mention.

In addition to these, Special Prizes could have been awarded for particularly elegant solutions, which the Jury considered outstanding. These were rare cases; in the period covered by this book, a Special Prize was awarded only in 1986 and 1995.

27th IMO (1986) in Poland

Country	Score	Prizes		
		First	Second	Third
U.S.S.R.	203	2	4	—
U.S.A.	203	3	3	—
F.R. Germany	196	2	4	—
China	177	3	1	1
German D.R.	172	1	3	2
Romania	171	2	2	1
Bulgaria	161	1	3	2
Hungary	151	1	2	2
Czechoslovakia	149	—	3	3
Vietnam	146	1	2	2
United Kingdom	141	—	2	3
France	131	1	1	2
Austria	127	—	2	2
Israel	119	—	2	2
Australia	117	—	—	5
Canada	112	—	2	1
Poland	93	—	—	3
Morocco	90	—	1	2
Tunisia	85	—	—	1
Yugoslavia	84	—	—	2
Algeria	80	—	—	2
Belgium	79	—	1	2
Spain (4)	78	—	1	2
Brazil	69	1	—	—
Norway	68	—	1	—
Greece	63	—	—	2
Finland	60	—	—	1
Colombia	58	—	—	—
Sweden	57	—	—	1
Turkey	55	—	—	—
Mongolia	54	—	—	—
Cyprus	53	—	1	—
Cuba	51	—	—	—
Italy (3)	49	—	—	2
Kuwait (5)	48	—	—	—
Iceland (4)	37	—	—	—
Luxembourg (2)	22	—	—	—

Special Prize for the solution of problem 1986/3: Joseph Keane (USA).

28th IMO (1987) in Cuba

		Prizes		
Country	Score	First	Second	Third
Romania	250	5	1	—
F.R. Germany	248	4	2	—
U.S.S.R.	235	3	3	—
German D.R.	231	2	3	1
U.S.A.	220	2	3	1
Hungary	218	—	5	1
Bulgaria	210	1	3	2
China	200	2	2	2
Czechoslovakia	192	—	4	2
United Kingdom	182	1	2	2
Vietnam	172	—	1	5
France	154	—	3	2
Austria	150	—	2	3
Netherlands	146	—	1	4
Australia	143	—	3	—
Canada	139	1	1	1
Sweden	134	—	2	2
Yugoslavia	132	—	1	3
Brazil	116	1	—	2
Greece	111	—	—	4
Turkey	94	—	—	2
Spain	91	—	—	3
Morocco	88	—	—	3
Cuba	83	—	—	2
Belgium	74	—	—	1
Iran	70	—	—	1
Finland	69	—	—	2
Norway	69	—	—	—
Colombia	68	—	—	1
Mongolia	67	—	—	—
Poland (3)	55	—	—	2
Iceland (4)	45	—	—	—
Cyprus	42	—	—	—
Peru	41	—	—	—
Italy (4)	35	—	—	1
Luxembourg (1)	27	—	—	1
Algeria	29	—	—	—
Kuwait	28	—	—	—
Uruguay (4)	27	—	—	—
Mexico (5)	17	—	—	—
Nicaragua	13	—	—	—
Panama	7	—	—	—

29th IMO (1988) in Australia

		Medals			
Country	Score	Gold	Silver	Bronze	H. Men.
U.S.S.R.	217	4	2	—	—
Romania	201	2	4	—	—
China	201	2	4	—	—
F.R. Germany	174	1	4	1	—
Vietnam	166	1	4	—	—
U.S.A.	153	—	5	1	—
German D.R. (5)	145	1	4	—	—
Bulgaria	144	—	4	2	—
France	128	1	1	3	1
Canada	124	1	1	2	1
United Kingdom	121	—	3	2	—
Czechoslovakia	120	—	2	2	1
Sweden	115	1	—	4	1
Israel	115	1	—	4	1
Austria	110	1	1	1	1
Hungary	109	—	2	2	1
Australia	100	1	—	1	1
Singapore	96	—	2	2	1
Yugoslavia	92	—	—	4	1
Iran	86	—	1	3	—
Netherlands	85	—	—	3	1
South Korea	79	—	—	3	—
Belgium	76	—	—	3	1
Hong Kong	68	—	—	2	1
Tunisia (4)	67	—	—	3	—
Colombia	66	—	—	3	—
Turkey	65	—	—	3	—
Greece	65	—	—	1	3
Finland	65	—	—	2	1
Luxembourg (3)	64	—	1	2	—
Morocco	62	—	—	2	1
Peru	55	—	—	1	3
Poland (3)	54	—	1	—	2
New Zealand	47	—	1	—	—
Italy (4)	44	—	—	1	1
Algeria (5)	42	—	1	—	1
Mexico	40	—	—	1	2
Brazil	39	—	—	—	2
Iceland (4)	37	—	—	1	—
Cuba	35	—	—	—	1

29th IMO (1988) in Australia (continued)

		Medals			
Country	Score	Gold	Silver	Bronze	H. Men.
Spain	34	—	—	—	1
Norway	33	—	—	—	—
Ireland	30	—	—	—	—
Philippines (5)	29	—	—	—	1
Kuwait	23	—	—	—	—
Argentina (3)	23	—	—	—	1
Cyprus	21	—	—	—	1
Indonesia (3)	6	—	—	—	—
Ecuador (1)	1	—	—	—	—

30th IMO (1989) in Germany

		Medals			
Country	Score	Gold	Silver	Bronze	H. Men.
China	237	4	2	—	—
Romania	223	2	4	—	—
U.S.S.R.	217	3	2	1	—
German D.R.	216	3	2	1	—
U.S.A.	207	1	4	1	—
Czechoslovakia	202	2	1	3	—
Bulgaria	195	1	3	2	—
F.R. Germany	187	1	3	2	—
Vietnam	183	2	1	3	—
Hungary	175	—	4	1	1
Yugoslavia	170	1	3	1	1
Poland	157	—	3	3	—
France	156	—	1	5	—
Iran	147	—	2	3	1
Singapore	143	—	—	4	2
Turkey	133	—	1	4	1
Hong Kong	127	—	2	1	1
Italy	124	—	1	2	3
Canada	123	—	1	3	2
Greece	122	—	1	3	2
United Kingdom	122	—	2	1	2
Australia	119	—	2	2	—
Colombia	119	—	1	2	3

(continued)

30th IMO (1989) in Germany (continued)

		Medals			
Country	Score	Gold	Silver	Bronze	H. Men.
Austria	111	—	2	1	1
India	107	—	—	4	1
Israel	105	—	2	1	—
Belgium	104	—	—	3	2
South Korea	97	—	1	—	4
Netherlands	92	—	1	1	2
Tunisia	81	—	1	—	2
Mexico	79	—	—	1	3
Sweden	73	—	—	2	1
New Zealand	69	—	—	2	2
Cuba	69	—	—	1	3
Luxembourg (3)	65	—	1	1	—
Brazil	64	—	—	3	—
Norway (4)	64	—	—	1	2
Morocco	63	—	—	1	3
Spain	61	—	—	1	4
Finland	58	—	—	—	3
Thailand	54	—	—	1	2
Peru	51	—	—	—	3
Philippines	45	—	1	—	—
Portugal	39	—	—	—	4
Ireland	37	—	—	—	2
Iceland (4)	33	—	—	—	2
Kuwait	31	—	—	—	—
Cyprus	24	—	—	—	1
Indonesia	21	—	—	—	—
Venezuela (4)	6	—	—	—	—

31st IMO (1990) in China

		Medals			
Country	Score	Gold	Silver	Bronze	H. Men.
China	230	5	1	—	—
U.S.S.R.	193	3	2	1	—
U.S.A.	174	2	3	—	—
Romania	171	2	2	2	—
France	168	3	1	—	—

31st IMO (1990) in China (continued)

Country	Score	Medals			
		Gold	Silver	Bronze	H. Men.
Hungary	162	1	3	2	—
German D.R.	158	—	4	2	—
Czechoslovakia	153	—	5	1	—
Bulgaria	152	1	4	1	—
United Kingdom	141	2	—	2	1
Canada	139	—	3	1	2
F.R. Germany	138	—	2	4	—
Italy	131	1	1	4	—
Iran	122	—	4	—	—
Australia	121	—	2	4	—
Austria	121	—	1	4	—
India	116	1	1	2	—
Norway	112	—	3	1	—
North Korea	109	—	1	3	—
Japan	107	—	2	1	—
Poland	106	—	2	1	2
Hong Kong	105	—	—	4	1
Vietnam	104	—	1	3	—
Brazil	102	1	—	2	—
Yugoslavia	98	—	1	2	1
Israel	95	—	1	3	—
Singapore	93	—	—	2	2
Sweden	91	—	1	2	—
Netherlands	90	—	1	2	2
Colombia	88	—	1	2	—
New Zealand	83	—	—	2	2
South Korea	79	—	1	1	1
Thailand	75	—	—	2	1
Turkey	75	—	—	1	2
Spain	72	—	—	—	—
Morocco (5)	71	—	1	—	—
Mexico	69	—	—	1	2
Argentina	67	—	—	1	2
Cuba	67	—	—	1	1
Ireland	65	—	—	1	—
Bahrain	65	—	—	—	1
Greece	62	—	—	1	1
Finland	59	—	—	1	1
Luxembourg (2)	58	1	—	1	—
Tunisia (4)	55	—	1	1	—

(continued)

31st IMO (1990) in China (continued)

Country	Score	Medals			
		Gold	Silver	Bronze	H. Men.
Mongolia	54	—	—	—	3
Kuwait (4)	53	—	—	1	1
Cyprus (4)	46	—	—	1	—
Philippines	46	—	—	1	—
Portugal	44	—	—	—	—
Indonesia	40	—	—	—	—
Macau	32	—	—	—	—
Iceland (3)	30	—	—	1	—
Algeria (4)	29	—	—	—	—

32nd IMO (1991) in Sweden

Country	Score	Medals			
		Gold	Silver	Bronze	H. Men.
U.S.S.R.	241	4	2	—	—
China	231	4	2	—	—
Romania	225	3	2	1	—
Germany	222	1	5	—	—
U.S.A.	212	1	4	1	—
Hungary	209	2	3	1	—
Bulgaria	192	—	3	3	—
Iran	191	2	1	2	—
Vietnam	191	—	4	2	—
India	187	—	3	3	—
Czechoslovakia	186	—	4	1	1
Japan	180	—	3	3	—
France	175	1	1	4	—
Canada	164	1	2	2	1
Poland	161	—	2	4	—
Yugoslavia	160	—	2	3	1
South Korea	151	—	1	4	1
Austria	142	—	2	3	—
United Kingdom	142	1	—	2	3
Australia	129	—	—	3	2
Sweden	125	—	2	1	—
Belgium	121	—	—	3	3
Israel	115	—	1	2	1
Turkey	111	—	—	4	1

32nd IMO (1991) in Sweden (continued)

		Medals			
Country	Score	Gold	Silver	Bronze	H. Men.
Thailand	103	—	1	1	3
Colombia	96	—	—	2	2
Argentina	94	—	—	3	1
Singapore	94	—	1	1	1
Hong Kong	91	—	—	2	1
New Zealand	91	—	—	2	1
Norway	85	—	—	3	—
Morocco	85	—	—	1	4
Greece	81	—	—	2	1
Cuba	80	—	—	2	2
Mexico	76	—	—	1	3
Italy	74	—	—	1	1
Netherlands	73	—	—	1	3
Brazil	73	—	—	1	1
Tunisia (4)	69	—	—	2	1
Finland	66	—	—	1	1
Spain	66	—	—	1	—
Philippines (4)	64	—	—	2	1
Denmark (5)	49	—	—	—	1
Ireland	47	—	—	—	—
Trinidad & Tobago (4)	46	—	—	—	3
Portugal	42	—	—	—	—
Mongolia	33	—	—	—	3
Luxembourg (2)	30	—	—	1	—
Indonesia	30	—	—	—	1
Switzerland (1)	29	—	—	1	—
Iceland	29	—	—	1	—
Cyprus (4)	25	—	—	1	—
Algeria	20	—	—	—	1
Macau	18	—	—	—	—
Bahrain	4	—	—	—	—

33rd IMO (1992) in Russia

		Medals			
Country	Score	Gold	Silver	Bronze	H. Men.
China	240	6	—	—	—
U.S.A.	181	3	3	—	—
Romania	177	2	2	2	—
Cwth. Ind. States	176	2	3	—	1
United Kingdom	168	2	2	2	—
Russia	158	2	2	2	—
Germany	149	—	4	2	—
Hungary	142	1	3	1	—
Japan	142	1	3	1	1
France	139	1	3	1	—
Vietnam	139	1	2	3	—
Yugoslavia	136	—	2	4	—
Czechoslovakia	134	—	2	3	—
Iran	133	—	3	2	1
Bulgaria	127	1	1	3	—
North Korea	126	—	3	2	—
Taiwan	124	—	3	2	—
South Korea	122	1	—	4	—
Australia	118	1	1	2	1
Israel	108	—	2	2	1
India	107	—	1	4	1
Canada	105	1	—	3	1
Belgium	100	—	1	2	1
Sweden	90	—	2	—	—
Poland	90	—	1	3	1
Singapore	89	—	1	3	—
Hong Kong	89	—	1	2	2
Italy	83	—	—	3	1
Norway	77	—	1	2	1
Netherlands	71	—	1	—	2
Austria	70	—	—	3	—
Argentina	67	—	1	1	1
Tunisia (4)	64	1	—	1	—
Turkey	63	—	—	2	—
Colombia	55	—	—	1	—
Mongolia	51	—	—	—	2
Thailand	50	—	1	—	—
Spain	50	—	—	1	1
Brazil	48	—	—	1	1
Morocco	45	—	—	—	2

33rd IMO (1992) in Russia (continued)

Country	Score	Medals			
		Gold	Silver	Bronze	H. Men.
Denmark (5)	42	—	—	—	—
Ireland	42	—	—	—	1
New Zealand	41	—	—	1	—
Philippines (4)	40	—	—	1	1
Greece	37	—	—	—	2
Portugal	35	—	—	1	—
Macau	35	—	—	—	—
Cyprus	34	—	—	1	—
Finland	33	—	—	—	—
Mexico	32	—	—	—	2
Switzerland (3)	30	—	—	—	1
Trinidad & Tobago	26	—	—	—	—
Indonesia	22	—	—	—	—
South Africa	21	—	—	—	—
Cuba (3)	17	—	—	—	1
Iceland (3)	16	—	—	—	1

34th IMO (1993) in Turkey

Country	Score	Medals			
		Gold	Silver	Bronze	H. Men.
China	215	6	—	—	—
Germany	189	4	2	—	—
Bulgaria	178	2	4	—	—
Russia	177	4	1	1	—
Taiwan	162	1	4	1	—
Iran	153	2	3	1	—
U.S.A.	151	2	2	2	—
Hungary	143	3	1	2	—
Vietnam	138	1	4	1	—
Czech Republic	132	1	2	3	—
Romania	128	1	2	3	—
Slovakia	126	1	3	1	—
Australia	125	1	2	3	—
United Kingdom	118	—	3	3	—
India	116	—	4	1	—
South Korea	116	—	3	3	—
France	115	2	1	1	—

(continued)

34th IMO (1993) in Turkey (continued)

Country	Score	Medals			
		Gold	Silver	Bronze	H. Men.
Israel	113	1	2	2	—
Canada	113	1	1	3	—
Japan	98	—	2	3	—
Ukraine	96	—	2	3	—
Austria	87	—	1	4	—
Italy	86	1	—	2	—
Turkey	81	—	1	2	—
Kazakhstan	80	—	1	3	—
Georgia	79	—	1	3	1
Colombia	79	—	—	4	—
Armenia	78	1	1	—	—
Poland	78	—	2	1	1
Singapore	75	—	1	3	—
Latvia	73	—	2	1	1
Denmark	72	—	1	3	—
Hong Kong	70	—	—	4	1
Brazil	60	—	—	1	2
Netherlands	58	—	—	1	1
Cuba	56	—	1	1	1
Belgium	55	—	—	1	1
Belarus (4)	54	—	1	1	—
Sweden	51	—	1	1	—
Morocco	49	—	—	1	2
Thailand	47	—	—	2	—
Argentina	46	—	1	1	—
Switzerland (4)	46	—	1	1	—
Norway	44	—	—	2	—
Spain	43	—	1	1	—
New Zealand	43	—	—	2	—
Slovenia	43	—	—	2	1
Macedonia (4)	42	—	—	3	—
Lithuania	41	—	—	—	—
Ireland	39	—	—	1	—
Portugal	35	—	—	1	—
Azerbaijan	33	—	—	1	—
Philippines	33	—	—	1	—
Finland	33	—	—	—	—
Croatia	32	—	—	1	—
Estonia	31	—	—	1	—
South Africa	30	—	—	—	—

34th IMO (1993) in Turkey (continued)

		Medals			
Country	Score	Gold	Silver	Bronze	H. Men.
Trinidad & Tobago	30	—	—	—	—
Moldova	29	—	—	—	—
Kyrgyzstan	28	—	—	—	—
Mongolia	26	—	—	1	—
Mexico	24	—	—	1	—
Macau	24	—	—	—	—
Iceland (4)	23	—	—	—	—
Luxembourg (1)	20	—	1	—	—
Albania	18	—	—	—	—
T.R. Northern Cyprus	17	—	—	—	—
Bahrain	16	—	—	—	—
Kuwait	16	—	—	—	—
Indonesia	15	—	—	—	—
Bosnia & Herzegovina (2)	14	—	—	1	—
Algeria	9	—	—	—	—
Turkmenistan (3)	9	—	—	—	—

35th IMO (1994) in Hong Kong

		Medals			
Country	Score	Gold	Silver	Bronze	H. Men.
U.S.A.	252	6	—	—	—
China	229	3	3	—	—
Russia	224	3	2	1	—
Bulgaria	223	3	2	1	—
Hungary	221	1	5	—	—
Vietnam	207	1	5	—	—
United Kingdom	206	2	2	2	—
Iran	203	2	2	2	—
Romania	198	—	5	1	—
Japan	180	1	2	3	—
Germany	175	1	2	3	—
Australia	173	—	2	3	1
Poland	170	2	—	3	1
Taiwan	170	—	4	1	1
South Korea	170	—	2	4	—
India	168	—	3	3	—
Ukraine	163	1	1	2	2

(continued)

35th IMO (1994) in Hong Kong (continued)

Country	Score	Medals			
		Gold	Silver	Bronze	H. Men.
Hong Kong	162	—	2	4	—
France	161	1	1	3	—
Argentina	159	—	3	1	—
Czech Republic	154	—	2	2	2
Slovakia	150	1	1	2	1
Belarus	144	—	1	4	1
Canada	143	1	—	3	1
Israel	143	—	1	4	1
Colombia	136	—	2	2	2
South Africa	120	—	—	3	1
Turkey	118	—	—	4	2
Singapore	116	—	2	—	3
New Zealand	116	—	—	4	1
Austria	114	1	—	—	2
Armenia (5)	110	—	—	4	1
Thailand	106	—	—	3	1
Belgium	105	—	—	2	4
Morocco	105	—	—	2	4
Italy	102	—	—	2	3
Netherlands	99	—	—	2	3
Latvia	98	—	—	3	1
Brazil (5)	95	—	2	—	3
Georgia	95	—	—	2	3
Sweden	92	—	—	1	3
Greece	91	—	—	1	5
Croatia	90	—	—	2	2
Estonia (5)	82	—	—	1	3
Norway	80	—	1	1	—
Macau	75	—	1	—	3
Lithuania	73	—	—	1	1
Finland	70	—	—	—	4
Ireland	68	—	—	—	3
Macedonia (4)	67	—	—	1	2
Mongolia	65	—	1	—	3
Trinidad & Tobago	63	—	—	—	1
Philippines	53	—	—	—	1
Chile (2)	52	—	1	—	1
Moldova	52	—	—	1	1
Portugal	52	—	—	—	—
Denmark (4)	51	—	—	2	—

35th IMO (1994) in Hong Kong (continued)

Country	Score	Medals			
		Gold	Silver	Bronze	H. Men.
Cyprus	48	—	—	—	1
Slovenia (5)	47	—	—	—	3
Indonesia	46	—	—	—	3
Bosnia & Herzegovina (5)	44	—	—	1	1
Spain	41	—	—	—	2
Switzerland (3)	35	—	—	1	—
Luxembourg (1)	32	—	1	—	—
Iceland (4)	29	—	—	—	1
Mexico	29	—	—	—	1
Kyrgyzstan	24	—	—	—	1
Cuba (1)	12	—	—	—	1
Kuwait (5)	12	—	—	—	—

36th IMO (1995) in Canada

Country	Score	Medals			
		Gold	Silver	Bronze	H. Men.
China	236	4	2	—	—
Romania	230	4	2	—	—
Russia	227	4	2	—	—
Vietnam	220	2	4	—	—
Hungary	210	3	1	2	—
Bulgaria	207	1	4	1	—
South Korea	203	2	3	1	—
Iran	202	2	3	1	—
Japan	183	1	3	2	—
United Kingdom	180	2	1	3	—
U.S.A.	178	—	3	3	—
Taiwan	176	—	4	1	1
Israel	171	1	2	2	1
India	165	—	3	3	—
Germany	162	1	3	1	1
Poland	161	—	1	5	—
Czech Republic	154	—	1	5	—
Yugoslavia	154	—	2	3	1
Canada	153	—	2	3	1
Hong Kong	151	—	2	3	1
Australia	145	—	1	4	1

(continued)

36th IMO (1995) in Canada (continued)

Country	Score	Medals			
		Gold	Silver	Bronze	H. Men.
Slovakia	145	—	2	2	2
Ukraine	140	1	1	1	2
Morocco	138	—	1	4	1
Turkey	134	—	2	3	—
Italy	131	—	—	5	1
Singapore	131	—	2	2	1
Belarus	131	—	1	3	2
Argentina	129	—	2	2	1
France	119	1	—	2	3
Macedonia	117	—	1	3	—
Armenia	111	—	2	1	1
Croatia	111	—	—	3	2
Thailand	107	—	1	2	1
Sweden	106	—	—	2	3
Finland	101	—	—	3	2
Moldova	101	—	1	1	2
Colombia	100	—	1	2	2
Switzerland (5)	97	—	2	—	2
Latvia	97	—	1	1	2
South Africa	95	—	—	2	4
Mongolia	91	—	—	1	5
Austria	88	—	—	1	3
Brazil	86	1	—	—	3
Netherlands	85	—	—	2	1
New Zealand	84	—	1	1	1
Belgium	83	—	—	1	4
Georgia	79	—	1	—	2
Denmark	77	—	—	1	4
Lithuania	74	—	—	—	4
Spain	72	—	—	1	3
Norway	70	—	—	1	2
Indonesia	68	—	—	1	3
Greece	66	—	—	1	2
Cuba (4)	59	—	—	—	3
Estonia	55	—	—	—	2
Kazakhstan	54	—	—	—	3
Mexico	43	—	—	1	—
Cyprus	43	—	—	—	3
Slovenia (5)	42	—	—	—	3
Ireland	41	—	—	—	2

36th IMO (1995) in Canada (continued)

		Medals			
Country	Score	Gold	Silver	Bronze	H. Men.
Macau	33	—	—	—	—
Trinidad & Tobago	32	—	—	—	2
Azerbaijan (3)	30	—	—	—	1
Philippines	28	—	—	1	—
Kyrgyzstan	28	—	—	—	1
Portugal	26	—	—	—	—
Iceland (4)	19	—	—	—	—
Bosnia & Herzegovina	18	—	—	—	—
Chile (2)	14	—	—	—	1
Sri Lanka (1)	10	—	—	—	1
Malaysia (2)	1	—	—	—	—
Kuwait (2)	0	—	—	—	—

Special Prize for the solution of problem 1995/6: Nikolay Nikolov (Bulgaria).

37th IMO (1996) in India

		Medals			
Country	Score	Gold	Silver	Bronze	H. Men.
Romania	187	4	2	—	—
U.S.A.	185	4	2	—	—
Hungary	167	3	2	1	—
Russia	162	2	3	1	—
United Kingdom	161	2	4	—	—
China	160	3	2	1	—
Vietnam	155	3	1	1	—
South Korea	151	2	3	—	—
Iran	143	1	4	1	—
Germany	137	3	1	1	—
Bulgaria	136	1	4	1	—
Japan	136	1	3	1	—
Poland	122	—	3	3	—
India	118	1	3	1	—
Israel	114	1	2	2	—
Canada	111	—	3	3	—
Slovakia	108	—	2	4	—
Ukraine	105	1	—	5	—
Turkey	104	—	2	3	—

(continued)

37th IMO (1996) in India (continued)

Country	Score	Medals			
		Gold	Silver	Bronze	H. Men.
Taiwan (China-Taipei)	100	—	2	3	1
Belarus	99	1	1	2	1
Greece	95	—	1	5	—
Australia	93	—	2	3	—
Yugoslavia	87	—	1	2	1
Singapore	86	1	—	3	—
Italy	86	—	2	2	—
Hong Kong	84	—	1	4	—
Czech Republic	83	—	2	1	—
Argentina	80	—	1	3	—
Georgia	78	1	—	2	1
Belgium	75	—	—	4	—
Lithuania	68	—	1	2	1
Latvia	66	—	—	3	1
Croatia	63	—	1	1	—
Armenia	63	—	—	1	3
France	61	—	2	—	—
New Zealand	60	—	—	3	—
Norway	60	—	—	3	—
Finland	58	—	—	2	2
Sweden	57	—	1	1	1
Moldova (5)	55	—	—	2	1
Austria	54	—	1	—	—
South Africa	50	—	—	2	—
Mongolia	49	—	—	2	1
Slovenia	49	—	—	2	—
Colombia	48	—	1	—	—
Thailand	47	—	—	1	2
Denmark	44	—	—	2	—
Macedonia	44	—	—	2	—
Macau	44	—	—	1	1
Spain	44	—	—	—	1
Brazil	36	—	—	—	1
Sri Lanka	34	—	—	1	—
Mexico	34	—	—	—	—
Estonia	33	—	—	—	—
Iceland	31	—	—	1	—
Bosnia & Herzegovina (4)	30	—	—	1	—
Azerbaijan	27	—	—	—	1
Netherlands	26	—	—	—	—

37th IMO (1996) in India (continued)

Country	Score	Medals			
		Gold	Silver	Bronze	H. Men.
Trinidad & Tobago	25	—	—	—	—
Ireland	24	—	—	—	—
Switzerland (4)	23	—	—	1	—
Portugal	21	—	—	—	1
Kazakhstan	20	—	—	—	—
Morocco	19	—	—	1	—
Cuba (1)	16	—	—	1	—
Albania (4)	15	—	—	—	—
Kyrgyzstan	15	—	—	—	—
Cyprus (5)	14	—	—	—	—
Indonesia	11	—	—	—	—
Chile (2)	10	—	—	—	—
Malaysia (4)	9	—	—	—	1
Turkmenistan (4)	9	—	—	—	—
Philippines	8	—	—	—	—
Kuwait (2)	1	—	—	—	—

38th IMO (1997) in Argentina

Country	Score	Medals			
		Gold	Silver	Bronze	H. Men.
China	223	6	—	—	—
Hungary	219	4	2	—	—
Iran	217	4	2	—	—
Russia	202	3	2	1	—
U.S.A.	202	2	4	—	—
Ukraine	195	3	3	—	—
Bulgaria	191	2	3	1	—
Romania	191	2	3	1	—
Australia	187	2	3	1	—
Vietnam	183	1	5	—	—
South Korea	164	1	4	1	—
Japan	163	1	3	1	1
Germany	161	1	3	2	—
Taiwan	148	—	4	2	—
India	146	—	3	3	—
United Kingdom	144	1	2	2	—
Belarus	140	—	2	4	—

(continued)

38th IMO (1997) in Argentina (continued)

Country	Score	Medals			
		Gold	Silver	Bronze	H. Men.
Czech Republic	139	1	2	2	—
Sweden	128	1	—	3	—
Yugoslavia	125	—	2	3	—
Poland	125	—	2	2	2
Israel	124	—	1	5	—
Latvia	124	—	1	4	1
Croatia	121	—	1	4	1
Turkey	119	—	1	4	—
Brazil	117	—	1	4	1
Colombia	112	—	—	6	—
Georgia	109	—	1	3	2
Canada	107	—	2	2	1
Mongolia	106	1	—	3	1
Hong Kong	106	—	—	5	—
France	105	1	—	1	2
Mexico	105	—	1	3	1
Finland	97	—	—	4	1
Armenia	97	—	—	3	1
Slovakia	96	—	1	2	1
Netherlands	94	—	2	—	2
Argentina	94	—	—	3	2
South Africa	93	1	—	2	1
Cuba	91	—	1	2	3
Singapore	88	—	—	4	1
Belgium	88	—	—	3	—
Austria	86	1	—	1	2
Norway	79	—	—	3	—
Greece	75	—	1	—	3
Macedonia	73	—	—	3	1
Kazakhstan	73	—	—	1	4
New Zealand	71	—	—	2	1
Italy	71	—	—	1	3
Slovenia	70	—	—	2	2
Lithuania	67	—	1	1	1
Thailand	66	—	—	1	4
Estonia	64	—	—	2	1
Peru	64	—	—	2	3
Azerbaijan	56	—	—	1	1
Macau	55	—	—	—	5
Moldova (3)	53	—	—	2	1

38th IMO (1997) in Argentina (continued)

Country	Score	Medals			
		Gold	Silver	Bronze	H. Men.
Switzerland (5)	53	—	—	2	—
Denmark	53	—	—	1	2
Iceland	48	—	1	—	—
Morocco	48	—	—	—	2
Bosnia & Herzegovina (5)	45	—	—	1	3
Indonesia	44	—	—	—	3
Spain	39	—	—	—	—
Trinidad & Tobago	30	—	—	—	1
Chile	28	—	—	—	2
Uzbekistan (3)	23	—	—	—	2
Ireland	21	—	—	—	—
Malaysia	19	—	—	—	1
Uruguay	19	—	—	—	1
Albania (3)	15	—	—	—	2
Portugal (5)	15	—	—	—	—
Philippines (2)	14	—	—	—	1
Bolivia (3)	13	—	—	—	1
Kyrgyzstan (3)	11	—	—	—	—
Kuwait (4)	8	—	—	—	—
Paraguay	8	—	—	—	—
Puerto Rico	8	—	—	—	—
Guatemala	7	—	—	—	—
Cyprus (3)	5	—	—	—	—
Venezuela (3)	4	—	—	—	—
Algeria (4)	3	—	—	—	—

39th IMO (1998) in Taiwan

Country	Score	Medals			
		Gold	Silver	Bronze	H. Men.
Iran	211	5	1	—	—
Hungary	186	4	2	—	—
Bulgaria	195	3	3	—	—
U.S.A.	186	3	3	—	—
Taiwan	184	3	2	1	—
Russia	175	2	3	1	—
India	174	3	3	—	—
Ukraine	166	1	3	2	—

(continued)

39th IMO (1998) in Taiwan (continued)

Country	Score	Medals			
		Gold	Silver	Bronze	H. Men.
Vietnam	158	1	3	2	—
Yugoslavia	156	—	5	—	1
Romania	155	3	—	2	1
South Korea	154	2	2	2	—
Australia	146	—	4	2	—
Japan	139	1	1	3	1
Czech Republic	135	—	3	3	—
Germany	129	—	3	2	—
Turkey	122	—	2	4	—
United Kingdom	122	—	1	4	1
Belarus	118	—	1	4	—
Canada	113	1	1	2	1
Poland	112	1	1	1	3
Singapore	110	—	1	3	2
Croatia	110	—	—	5	—
Israel	104	—	—	5	—
Hong Kong	102	—	1	3	1
France	100	1	—	2	2
Armenia	100	—	2	2	—
South Africa	98	—	1	2	3
Argentina	97	1	—	3	—
Brazil	91	1	—	1	2
Mongolia	91	—	2	2	—
Greece	90	—	2	1	1
Slovakia	88	—	1	4	—
Bosnia & Herzegovina	88	—	1	2	3
Kazakhstan	81	—	—	2	3
Georgia	78	—	—	3	2
Latvia	74	—	1	3	—
Italy	72	—	—	3	2
Belgium	71	—	1	1	—
Macedonia	69	—	—	1	1
Colombia	66	1	—	—	2
Thailand	65	—	—	2	1
Estonia	63	—	1	1	—
Mexico	62	—	1	—	1
Netherlands	62	—	1	—	—
Peru (3)	60	—	2	—	1
Sweden	58	—	—	2	—
Austria	57	—	—	2	1

39th IMO (1998) in Taiwan (continued)

Country	Score	Medals			
		Gold	Silver	Bronze	H. Men.
New Zealand	50	—	—	2	—
Moldova (2)	45	—	1	1	—
Slovenia	44	—	—	1	2
Iceland	42	—	—	—	3
Morocco	42	—	—	—	3
Azerbaijan (5)	41	—	—	1	1
Lithuania	40	—	—	1	1
Cyprus (4)	39	—	—	1	2
Switzerland	37	—	—	—	2
Spain	36	—	—	1	1
Ireland	36	—	—	1	—
Trinidad & Tobago	36	—	—	1	—
Norway	33	—	—	—	1
Malaysia	32	—	—	—	—
Finland	30	—	—	—	1
Macau (5)	29	—	—	—	2
Luxembourg (2)	25	—	—	1	1
Denmark	21	—	—	—	—
Cuba (1)	19	—	—	1	—
Indonesia (5)	16	—	—	—	—
Kyrgyzstan (5)	14	—	—	—	—
Philippines (4)	11	—	—	—	—
Uruguay	11	—	—	—	—
Paraguay (5)	6	—	—	—	—
Portugal	6	—	—	—	—
Sri Lanka (1)	5	—	—	—	—
Venezuela (2)	1	—	—	—	—
Kuwait (3)	0	—	—	—	—

40th IMO (1999) in Romania

Country	Score	Medals			
		Gold	Silver	Bronze	H. Men.
China	182	4	2	—	—
Russia	182	4	2	—	—
Vietnam	177	3	3	—	—
Romania	173	3	3	—	—
Bulgaria	170	3	3	—	—

(continued)

40th IMO (1999) in Romania (continued)

Country	Score	Medals			
		Gold	Silver	Bronze	H. Men.
Belarus	167	3	3	—	—
South Korea	164	3	3	—	—
Iran	159	2	4	—	—
Taiwan	153	1	5	—	—
U.S.A.	150	2	3	1	—
Hungary	147	1	4	1	—
Ukraine	136	2	2	1	—
Japan	135	2	4	—	—
Yugoslavia	130	1	2	3	—
Australia	116	1	1	3	1
Turkey	109	1	1	4	—
Germany	108	—	2	4	—
India	107	—	3	3	—
Poland	104	1	—	5	—
United Kingdom	100	—	3	2	—
Slovakia	88	—	2	3	—
Latvia	86	1	1	—	—
Italy	82	—	1	2	—
Switzerland	79	—	1	3	—
Mongolia	78	—	2	1	—
Israel	78	—	—	5	1
Cuba	77	—	1	4	—
South Africa	77	—	1	1	—
Austria	75	—	1	2	—
Brazil	75	—	—	4	—
Netherlands	74	—	—	4	—
Canada	74	—	—	3	—
France	73	—	1	2	1
Hong Kong	73	—	—	4	1
Kazakhstan	72	—	—	4	1
Macedonia	71	—	—	5	—
Singapore	71	—	—	4	—
Georgia	68	—	1	1	1
Norway	67	—	1	2	—
Armenia	67	—	—	3	—
Sweden	66	—	—	3	—
Croatia	66	—	—	2	—
Finland	65	—	1	—	1
Bosnia & Herzegovina	65	—	—	3	—
Argentina	63	—	—	3	—

40th IMO (1999) in Romania (continued)

		Medals			
Country	Score	Gold	Silver	Bronze	H. Men.
Spain	60	—	—	1	1
Greece	57	—	2	—	—
Thailand	57	—	—	3	—
Colombia	55	—	1	1	—
Czech Republic	55	—	—	1	1
Lithuania	54	—	—	2	—
Mexico	53	—	—	1	—
New Zealand	53	—	—	1	—
Denmark (5)	51	—	—	2	—
Belgium	51	—	—	2	—
Moldova	50	—	—	1	1
Morocco	48	—	—	1	1
Slovenia	46	—	—	2	—
Uzbekistan	42	—	—	—	—
Iceland	41	—	—	1	—
Macau	41	—	—	—	—
Ireland	38	—	—	1	—
Malaysia	37	—	—	—	—
Cyprus	35	—	—	—	1
Indonesia	35	—	—	—	—
Azerbaijan	34	—	—	1	—
Albania (5)	34	—	—	—	—
Trinidad & Tobago	33	—	—	—	—
Estonia (4)	30	—	—	1	—
Portugal	29	—	—	—	—
Luxembourg (2)	26	—	—	1	—
Uruguay (5)	25	—	—	—	—
Philippines (4)	24	—	—	—	—
Tunisia (4)	22	—	—	—	—
Guatemala	19	—	—	—	—
Kyrgyzstan (5)	15	—	—	—	—
Turkmenistan (2)	13	—	—	—	—
Peru (2)	10	—	—	—	—
Kuwait (4)	10	—	—	—	—
Venezuela (2)	8	—	—	—	—
Sri Lanka (1)	6	—	—	—	—

List of Symbols

$P \Longrightarrow Q$	statement P implies statement Q
$P \Longleftrightarrow Q$	statements P and Q are equivalent
$x \in X$	x is an element of the set X
$X \cup Y$	the union of sets X and Y
$X \cap Y$	the intersection of sets X and Y
$X \setminus Y$	the set X without Y
$\{x \in X\colon\ \Pr(x)\}$	the set of all x in X that have property $\Pr(x)$
$\{x_1, \ldots, x_n\}$	the set whose elements are $x_1, \ldots, x_n$
$(x_1, \ldots, x_n)$	the ordered array (sequence, vector, n-tuple) whose consecutive terms are $x_1, \ldots, x_n$
$\lvert X \rvert$	the number of elements in a finite set X
$f\colon X \to Y$	the function f maps the set X into Y
$g \circ f$	the composition of functions f and g
f^n	the n-fold iterate of a function $f\colon X \to X$
R	the set of all real numbers
Z	the set of all integers
N_0	the set of all nonnegative integers
N or Z^+	the set of all positive integers (natural numbers)
$k \equiv \ell \pmod{m}$	the integers k and ℓ are congruent modulo m
$\sum_{i=1}^{n} x_i,\ \prod_{i=1}^{n} x_i$	the sum and the product of the numbers $x_1, \ldots, x_n$
$n!$	n factorial; the product of integers from 1 to n ($0! = 1$)

$\binom{n}{k}$	the binomial coefficient; $\binom{n}{k} = \dfrac{n!}{k!(n-k)!}$
$\lfloor x \rfloor$	the greatest integer not exceeding the real number x
$[a, b]$	the closed interval $\{x \in \mathrm{R}:\ a \le x \le b\}$
(a, b)	the open interval $\{x \in \mathrm{R}:\ a < x < b\}$
(x, y)	in geometry: the point with coordinates x, y
AB	in geometry: the segment with endpoints A and B; also the length of the segment AB
$\overrightarrow{AB}$	the vector AB, directed from A to B

Glossary of Frequently Used Terms and Theorems

Angle Two rays originating at the same point divide the plane into two parts, each part being called an *angle*. If the rays do not lie in a line, then one of the two angles is *convex* (the one with size $< 180°$) and the other one *concave* (size $> 180°$).

Angles in a circle If A, B, X are distinct points on a circle, the angle AXB is *subtended* by the arc AB not containing the point X; we also say that the angle is subtended by the chord AB, if it is clear from the context which one of the two arcs AB is the subtending one.

Two angles inscribed in the same circle and subtended by equal arcs have equal size. If AZ is the ray emanating from A and tangent to the arc AB subtending $\angle AXB$, then $\angle AXB = \angle BAZ$ (the *tangent-chord theorem*).

Area of a triangle Triangle ABC with sides a, b, c opposite to angles A, B, C and the respective altitudes h_a, h_b, h_c has area $F = \frac{1}{2}ah_a = \frac{1}{2}bh_b = \frac{1}{2}ch_c$ or $F = \frac{1}{2}bc\sin A = \frac{1}{2}ca\sin B = \frac{1}{2}ab\sin C$, or $F = \frac{1}{2}(a+b+c)r$, where r is the inradius.

Arithmetic mean–geometric mean inequality (AM-GM inequality)

$$\frac{a_1 + \cdots + a_n}{n} \geq (a_1 \cdots a_n)^{1/n},$$

holding for all real numbers $a_1, \ldots, a_n \geq 0$, with equality if and only if $a_1 = \cdots = a_n$.

Arithmetic progression Sequence $x_1, x_2, x_3, \ldots$ such that all the differences $x_{i+1} - x_i$ are equal; their common value is the *step* of the progression.

Bijective function (one-to-one mapping) Function which is injective and surjective.

Binary representation Representation of a positive integer n in the form

$$n = (c_k c_{k-1} \ldots c_1 c_0)_2 = c_k 2^k + c_{k-1} 2^{k-1} + \cdots + c_1 2^1 + c_0 2^0$$

with each c_i equal to either 0 or 1.

Binomial coefficients For integers $n \geq k \geq 0$,

$$\binom{n}{k} = \frac{n!}{k!(n-k)!} = \text{coefficient of } x^k \text{ in the expansion of } (x+1)^k$$
$$= \text{number of ways to choose } k \text{ objects out of } n.$$

Properties: $\binom{n}{0} = \binom{n}{n} = 1;\ \binom{n+1}{k+1} = \binom{n}{k+1} + \binom{n}{k}.$

If p is a prime and k an integer, $0 < k < p$, then $\binom{p}{k}$ is divisible by p.

Cardinality of a finite set The number of elements in the set.

Cauchy–Schwarz inequality

$$(x_1 y_1 + \cdots + x_n y_n)^2 \leq (x_1^2 + \cdots + x_n^2)(y_1^2 + \cdots + y_n^2),$$

holding for all real numbers $x_1, \ldots, x_n$ and $y_1, \ldots, y_n$.

If the n-tuples $(x_1, \ldots, x_n)$ and $(y_1, \ldots, y_n)$ are viewed as vectors in n-space, the inequality (after taking square roots on both sides) says that the absolute value of their dot product does not exceed the product of their lengths; see *Vectors*.

Ceva's theorem Let D, E, F be points on sides BC, CA, AB of a triangle ABC. The segments AD, BE, CF concur if and only if

$$\frac{BD}{DC} \cdot \frac{CE}{EA} \cdot \frac{AF}{FB} = 1;$$

equivalently, if and only if

$$\frac{\sin EBC}{\sin BCF} \cdot \frac{\sin FCA}{\sin CAD} \cdot \frac{\sin DAB}{\sin ABE} = 1.$$

Chebyshev's inequality See problem 1995/2, third solution.

Chinese remainder theorem Let $a_1, a_2, \ldots, a_n$ and $r_1, r_2, \ldots, r_n$ be given integers. If the numbers a_i are pairwise relatively prime, then there exists an integer N such that $N \equiv r_i \pmod{a_i}$ for $i = 1, \ldots, n$.

Circumcircle Circumscribed circle. Its center and radius are called the *circumcenter, circumradius.*

Collinear points Points lying on the same straight line.

Complex numbers Pairs (x, y) of real numbers, written as sums $x + iy$, with operations of addition and multiplication satisfying the commutative, associative and distributive laws, accompanied by $i^2 = -1$. A number of the form $(x, 0)$, i.e., $x + 0 \cdot i$, is identified with the real x.

Pairs of real numbers are naturally interpreted as points of the plane; thus the set of all complex numbers is often called the *complex plane.*

A complex number $z = x + iy$ can be represented in the trigonometric form $z = \rho(\cos\varphi + i\sin\varphi)$, where $\rho = \sqrt{x^2 + y^2}$; then *de Moivre's formula* $z^n = \rho^n(\cos n\varphi + i\sin n\varphi)$ holds for every natural exponent n.

Composition of functions If f and g are functions such that the range of f is contained in the domain of g, then the (well-defined) function $h(x) = g(f(x))$ is the *composition* of f and g; in symbols, $h = g \circ f$. Analogously, if $f_1, \ldots, f_n$ are such that the range of f_i is contained in the domain of f_{i+1}, then the composition $h = f_n \circ \cdots \circ f_1$ is defined by $h(x) = f_n(\ldots(f_1(x))\ldots)$. When all f_is are the same function f, then $f \circ \cdots \circ f$ is the n-fold *iterate* of f (denoted by f^n).

Congruence Integers k and ℓ are *congruent* modulo an integer $m > 0$ if the difference $k - \ell$ is divisible by m; in symbols: $k \equiv \ell \pmod{m}$.

Convex function Real-valued function (in some real-line interval) such that, for any two points P, Q on its graph, there is no point on the portion of the graph between P and Q that would lie above the line segment PQ. A function f is *concave* if $-f$ is convex.

If f has a derivative f' and if f' is a nondecreasing function, then f is convex. See also *Jensen's inequality.*

Cyclic polygon Polygon that can be inscribed in a circle. A convex quadrilateral $ABCD$ is cyclic if and only if $\angle A + \angle C = \angle B + \angle D\ (= 180°)$; equivalently, if $PA \cdot PC = PB \cdot PD$, where P is the point of intersection of diagonals. See also *Ptolemy's theorem.*

De Moivre's formula See *Complex numbers.*

Dot product of vectors See *Vectors* and *Cauchy–Schwarz inequality.*

Equipotent sets Sets of equal cardinalities. Equivalently: two sets are equipotent if there exists a bijective mapping of one set onto the other.

Excircle Escribed circle (of a triangle), tangent to one side of the triangle and to the lines (produced) containing the two other sides.

Fermat's little theorem If p is prime, then $a^p \equiv a \pmod{p}$ for every integer a. Hence, if a is nondivisible by p, then $a^{p-1} \equiv 1 \pmod{p}$.

Fixed point of a function f Point x such that $f(x) = x$.

Graph (in combinatorics) Set with a distinguished family of its two-element subsets (*edges*); see the statement of problem 1991/4 and the remark after the solution to problem 1992/3 for more details.

Graph of a function If f is a real-valued function on a subset A of the real-line, its *graph* is the set in the Cartesian plane consisting of all points $(x, f(x))$, $x \in A$.

Homothety Dilatation with center O and ratio λ, transformation of the plane sending each point X to its image position Y defined by the vector equality $\overrightarrow{OY} = \lambda \cdot \overrightarrow{OX}$. The image of any figure is similar to the given figure in ratio λ.

Incircle Inscribed circle. Its center and radius are called the *incenter*, *inradius*.

Inclusion-exclusion formula (or principle) See problem 1989/6, third solution, and the remark after the solution to problem 1991/3.

Injective function Function $f \colon A \to B$ such that $f(x_1) \neq f(x_2)$ if $x_1 \neq x_2$ $(x_1, x_2 \in A)$.

Inversion Inversion with respect to a circle with center O and radius r, transformation of the plane sending each point X other than O to its image position Y defined by the vector equality

$$\overrightarrow{OY} = \left(\frac{r}{OX}\right)^2 \cdot \overrightarrow{OX};$$

then also Y is mapped to X, and X, Y are *inverse points*.

It is purposeful to adjoin to the plane one additional element, called the *point at infinity*, and consider it as the point inverse to O.

The image of any circle not passing through O is also a circle. The image of a circle passing through O is a straight line (completed with the point at infinity).

Iterates See *Composition of functions*.

Jensen's inequality

$$f(\lambda_1 x_1 + \cdots + \lambda_n x_n) \le \lambda_1 f(x_1) + \cdots + \lambda_n f(x_n),$$

holding for every convex function f (in an interval I), any numbers $x_1, \ldots, x_n \in I$ and any $\lambda_1, \ldots, \lambda_n \ge 0$ with $\sum \lambda_i = 1$. If f is concave, the inequality is reversed.

Lattice points In the Cartesian plane (or space), points with integer coefficients.

Law of sines In a triangle with circumradius R and sides a, b, c opposite to angles A, B, C: $a = 2R \sin A$, $b = 2R \sin B$, $c = 2R \sin C$.

Locus The set of all points in the plane (or space) that satisfy a property under consideration.

Matrix Rectangular array of numbers:

$$\begin{bmatrix} a_{11} & \ldots & a_{1n} \\ \vdots & \ddots & \vdots \\ a_{m1} & \ldots & a_{mn} \end{bmatrix}.$$

The vector $(a_{i1}, \ldots, a_{in})$ is the ith *row* and the vector $(a_{1j}, \ldots, a_{mj})$ is the jth *column* of the matrix.

Natural numbers Positive integers.

Orthocenter of a triangle Point of intersection of altitudes.

Orthogonal projection of a point onto a line The foot of the perpendicular dropped from that point to the line.

Pascal's theorem If $A_1, A_2, A_3, A_4, A_5, A_6$ are distinct points on a circle and if the lines A_1A_3, A_2A_4 intersect at P, lines A_2A_5, A_3A_6 intersect at Q, lines A_4A_6, A_5A_1 intersect at R, then the points P, Q, R are collinear.

Permutation Bijective mapping of a finite set S onto itself. If the elements of S are arranged into a sequence $(x_1, \ldots, x_n)$, a permutation can be regarded as a rearrangement; notation $\pi = (y_1, \ldots, y_n)$ then means that the permutation π maps each x_i to the corresponding y_i.

Pigeonhole principle If n objects are distributed among k boxes, $k < n$, then some box contains at least two elements.

More generally, if n objects are distributed among k boxes, $km < n$ ($m \ge 1$ an integer), then some box contains at least $m + 1$ elements.

Polygon The figure (in the plane) bounded by a closed polygonal line, whose segments are called the *sides* of the polygon. A polygon with n sides is an n-gon.

NB. The term *polygon* is sometimes used for *polygonal line*.

Polygonal line Broken line, union of directed segments in succession, each one (starting from the second one) attached at the end of the preceding one. If also the first segment is attached at the end of the last one, the line is *closed*.

Polynomial Function of the form

$$P(x) = a_0 + a_1 x + a_2 x^2 + \cdots + a_n x^n.$$

Constants $a_0, \dots, a_n$ are the *coefficients*; a_n is the *leading coefficient* and a_0 is the *free term*; n is the *degree* of P, provided that $a_n \neq 0$.

Every number x_0 such that $P(x_0) = 0$ is a *root* of P. Every polynomial of degree $n \geq 1$, with real or complex coefficients, has roots in the complex numbers (the *fundamental theorem of algebra*) and factors uniquely into the product

$$P(x) = a_n (x - z_1)^{k_1} \cdots (x - z_r)^{k_r}$$

where $z_1, \dots, z_r$ are all distinct complex roots of P and $k_1, \dots, k_r$ are positive integers; k_j is the *multiplicity* of the root z_j. The sum of the exponents k_j is n, so the factorization formula can be rewritten as

$$P(x) = a_n (x - z_1) \cdots (x - z_n)$$

where now $z_1, \dots, z_n$ is the sequence of all complex roots of P, each root repeated as many times as its multiplicity indicates. The sum of all the roots (counting multiplicities) is equal to $-a_{n-1}/a_n$ and their product is $(-1)^n a_0/a_n$; these are two of *Viète's formulas*.

Power of a point If a line passing through a point P intersects a given circle ω at X and Y, then the product $PX \cdot PY$ has a constant value (depending on P and ω, but not on the choice of the line). This value, equipped with sign plus or minus according as P lies outside or inside the circle, is the *power* of P with respect to ω, and is equal to $OP^2 - r^2$, where r is the radius of ω.

Prime numbers Integers $p > 1$ that have no positive divisors other than 1 and p.

The fundamental property: if the product of two integers is divisible by a prime p, then at least one of the two factors is divisible by p.

Every integer greater than 1 is uniquely representable as the product of primes (the *prime factorization*).

Ptolemy's theorem Inequality $AC \cdot BD \leq AB \cdot CD + BC \cdot DA$, holding in every convex quadrilateral $ABCD$, with equality if and only if the quadrilateral is cyclic.

Radical axes The set of all points in the plane whose powers with respect to two given circles are equal (see *Power of a point*) is a straight line, called the *radical axis* (or *power axis*, or just *axis*) of the two circles. If the circles intersect, their axis is the line through the points of intersection.

The radical axes of three circles with noncollinear centers, taken in pairs, are concurrent.

Ray Half-line. Part of a straight line, consisting of all points on one side of a specified point on that line.

Recursion formula (recurrence formula) Formula that defines each entry of a sequence, except the initial one, in terms of the preceding one (ones). The initial entry has to be defined explicitly.

Relatively prime numbers (coprime numbers) Integers that have no common divisor greater than 1.

If n and m are relatively prime integers, then there exist integers u and v such that $nu + mv = 1$; when n and m are positive, it can be required that u be positive and v be negative (or vice versa).

Root of an equation Any number satisfying that equation; see *Polynomial*.

Root of a polynomial *P*, of multiplicity *k* Complex number z_0 such that $P(x)$ is divisible by $(x - z_0)^k$ but not by $(x - z_0)^{k+1}$; see *Polynomial*.

Roots of unity Solutions of the equation $z^n = 1$ (n natural); i.e., complex numbers of the form $\cos\varphi + i\sin\varphi$ where $\varphi = 2k\pi/n$, $k = 0, 1, \ldots, n-1$; see *Complex numbers*.

Surjective function Function $f\colon A \to B$ such that every $b \in B$ is the value $f(a)$ for some $a \in A$.

Transformation of the plane Mapping of the plane onto itself. Important types of transformations: translation (by a vector), rotation (about a point, by a specified angle), homothety (with a given center and ratio), inversion (with respect to a circle). If a transformation preserves the length and direction of every vector, then it is a translation. See *Homothety* and *Inversion*.

Vectors Directed segments, with the convention that if a directed segment is a translate of another one, they are both considered as representatives of the same vector. Notation: $\mathbf{v} = \overrightarrow{AB} (= \overrightarrow{CD}$ etc.). In n-dimensional space, a vector has n components: $\mathbf{v} = (x_1, \ldots, x_n)$; its length is equal to $\sqrt{x_1^2 + \cdots + x_n^2}$. Also in nongeometric considerations, finite sequences of numbers (n-tuples) are conveniently regarded as vectors.

The *sum* of vectors $\mathbf{v} = (x_1, \ldots, x_n)$ and $\mathbf{w} = (y_1, \ldots, y_n)$ is the vector $\mathbf{v} + \mathbf{w} = (x_1 + y_1, \ldots, x_n + y_n)$. Geometrically, if $\mathbf{v}$ and $\mathbf{w}$ are attached at the same point O so that $\mathbf{v} = \overrightarrow{OA}$, $\mathbf{w} = \overrightarrow{OB}$, then by completing the parallelogram $OASB$ we obtain $\mathbf{v} + \mathbf{w} = \overrightarrow{OS}$.

The *dot product* (also *scalar product, inner product*) of vectors $\mathbf{v}$ and $\mathbf{w}$ is the number $\mathbf{v} \cdot \mathbf{w} = x_1 y_1 + \cdots + x_n y_n$. Geometrically, this is the product of their lengths multiplied by the cosine of the angle between them; thus $\mathbf{v} \cdot \mathbf{w} = 0$ if and only if $\mathbf{v}$ and $\mathbf{w}$ are perpendicular. See *Cauchy–Schwarz inequality*.

Subject Classification

Many problems can be classified under more than one topic. In those cases, the primary classification is marked by an asterisk, and any further classification is indicated by the problem number in italics.

Number Theory

Divisibility, primes, factorization:

1986/1, 1987/6, 1988/6*, 1989/5, 1990/3, *1990/4*, *1990/5*, 1992/1, 1994/4, 1994/6*, *1995/6*, 1996/4, 1997/5, 1998/3*, 1998/4, 1998/6*, 1999/4

Integer lattice:

1987/5, *1988/6*

Binary representation:

1988/3*, *1994/3*, *1997/6*

Combinatorial arithmetic:

1987/4, 1990/6, 1991/2, *1991/3*, 1992/6, *1993/5*, *1994/1*, 1994/3*, *1994/6*, 1995/6*, *1996/3*, 1996/6, *1997/6*, *1998/3*

Combinatorics

Enumerative combinatorics:

1987/1, *1988/3*, *1989/3*, 1989/6, *1990/2*, 1991/3*, 1992/5, *1995/6*, 1997/6*, 1998/2, 1999/3

Partitioning, coloring:

1986/6, 1988/2, 1989/1, 1990/2*, *1992/3*, 1997/4*, *1997/6*

Graphs:

1991/4, 1992/3*, *1997/4*

Discrete procedures, games:

1986/3, 1990/5*, 1993/3, 1993/6, 1996/1

Combinatorial geometry:

1986/2, 1989/3*, 1995/3, 1997/1, 1999/1*

Algebra

Polynomials:

1993/1, *1995/6*

Functional equations:

1986/5, 1987/4*, 1990/4*, 1992/2, 1993/5*, 1994/5, 1996/3*, *1998/6*, 1999/6

Algebraic inequalities:

1987/3, 1988/4, 1994/1*, 1995/2, 1997/3*, 1999/2

Sequences:

1991/6, 1995/4, *1997/3*

Geometry

Concurrence, orthogonality, collinearity etc.:

1993/2, 1994/2, 1995/1, 1996/2, *1999/1*, 1999/5

Finding the locus:

1986/4, 1988/1, 1992/4

Calculating lengths and areas:

1986/2*, 1987/2, 1989/2*, 1990/1, 1993/2*, 1997/2, 1998/1

Geometric inequalities:

1988/5, *1989/2*, 1989/4, 1991/1, 1991/5, 1993/4, 1995/5, 1996/5, 1998/5

Bibliography

[1] Titu Andreescu, Kiran Kedlaya, Paul Zeitz, *Mathematical Contests 1995–1996. Olympiad Problems and Solutions from around the World*, American Mathematics Competitions, 1997.

[2] Titu Andreescu, Kiran Kedlaya, *Mathematical Contests 1996–1997. Olympiad Problems and Solutions from around the World*, American Mathematics Competitions, 1998.

[3] Titu Andreescu, Kiran Kedlaya, *Mathematical Contests 1997–1998. Olympiad Problems and Solutions from around the World*, American Mathematics Competitions, 1999.

[4] Titu Andreescu, Zuming Feng, *101 Problems in Algebra*, Enrichment Series, Australian Mathematics Trust, 2001.

[5] Titu Andreescu, Zuming Feng, *Mathematical Olympiads. Problems and Solutions from Around the World 1998–1999*, MAA, Washington, DC, 2000.

[6] Edward J. Barbeau, Murray S. Klamkin, William O. J. Moser, *Five Hundred Mathematical Challenges*, Spectrum Series, MAA, Washington, DC, 1995.

[7] Béla Bollobás, *Graph Theory, An Introductory Course*, Springer Verlag, NY, 1979.

[8] Oene Bottema, R. Ž. Djordjević, R. R. Janić, Dragoslav S. Mitrinović, Petar M. Vasić, *Geometric Inequalities*, Wolters-Noordhoff, Groningen, 1968.

[9] Harold S. M. Coxeter, Samuel L. Greitzer, *Geometry Revisited*, New Mathematical Library, Vol. 19, MAA, Washington, DC, 1967.

[10] Arthur Engel, *Problem-Solving Strategies*, Problem Books in Mathematics, Springer Verlag, NY, 1998.

[11] Dmitry Fomin, Alexey Kirichenko, *Leningrad Mathematical Olympiads 1997–1991*, MathPro Press, Mansfield, OH, 1994.

[12] Samuel L. Greitzer, *International Mathematical Olympiads 1959–1977*, New Mathematical Library, Vol. 27, MAA, Washington, DC, 1978.

[13] Paul R. Halmos, *Problems for Mathematicians, Young and Old*, Dolciani Mathematical Expositions, Vol. 12, MAA, Washington, DC, 1991.

[14] Ross Honsberger, *Mathematical Gems*, Dolciani Mathematical Expositions, Vols. I, II, IX, MAA, Washington, DC, 1973, 1976, 1985.

[15] Ross Honsberger, *Mathematical Morsels*, Dolciani Mathematical Expositions, Vol. III, MAA, Washington, DC, 1978.

[16] Ross Honsberger, *Mathematical Plums*, Dolciani Mathematical Expositions, Vol. IV, MAA, Washington, DC, 1979.

[17] Murray S. Klamkin, *International Mathematical Olympiads 1978–1985 and Forty Supplementary Problems*, New Mathematical Library, Vol. 31, MAA, Washington, DC, 1986.

[18] Murray S. Klamkin, *USA Mathematical Olympiads 1972–1986*, New Mathematical Library, Vol. 33, MAA, Washington, DC, 1988.

[19] Marcin E. Kuczma, *Problems. 144 Problems of the Austrian–Polish Mathematics Competition 1978–1993*, Academic Distribution Center, Freeland, MD, 1994.

[20] Marcin E. Kuczma, "On the 1997-IMO Problem Six," *Mathematics Competitions*, Australian Mathematics Trust, vol. 11/2 (1998) 45–57.

[21] Hans Lausch, *The Asian Pacific Mathematics Olympiad 1989–1993*, Enrichment Series, Australian Mathematics Trust, 1994.

[22] Hans Lausch, Peter J. Taylor, *Australian Mathematical Olympiads 1979–1995*, Enrichment Series, Australian Mathematics Trust, 1998.

[23] Andy Liu, *Chinese Mathematics Competitions and Olympiads 1981–1993*, Enrichment Series, Australian Mathematics Trust, 1998.

[24] Dragoslav S. Mitrinović, *Analytic Inequalities*, Springer Verlag, Heidelberg, 1970.

[25] Donald J. Newman, *A Problem Seminar*, Springer Verlag, NY, 1982.

[26] George Pólya, *How To Solve It*, Doubleday, NY, 1957.

[27] Igor F. Sharygin, *Problems in Plane Geometry*, Mir, Moscow, 1988.

[28] Igor F. Sharygin, *Problems in Solid Geometry*, Mir, Moscow, 1986.

[29] David O. Shklarsky, Nikolay N. Chentzov, I. M. Yaglom, *The USSR Olympiad Problem Book*, Freeman, 1962.

[30] David O. Shklarsky, Nikolay N. Chentzov, I. M. Yaglom, *Selected Problems and Theorems in Elementary Mathematics*, Mir, Moscow, 1979.

[31] Wacław Sierpiński, *250 Problems in Elementary Number Theory*, American Elsevier, NY, 1970.

[32] Arkadii M. Slinko, *USSR Mathematical Olympiads 1989–1992*, Enrichment Series, Australian Mathematics Trust, 1997.

[33] Peter J. Taylor, *Tournament of the Towns. Questions and Solutions 1984–1989*, Australian Mathematics Foundation, 1992.

[34] Peter J. Taylor, *Tournament of the Towns. Questions and Solutions 1989–1993*, Enrichment Series, Australian Mathematics Trust, 1994.

[35] Naum Y. Vilenkin, *Combinatorics*, Academic Press, NY, 1971.

[36] I. M. Yaglom, *Geometric Transformations I*, New Mathematical Library, Vol. 8, MAA, Washington, DC, 1962.

[37] I. M. Yaglom, *Geometric Transformations II*, New Mathematical Library, Vol. 21, MAA, Washington, DC, 1968.

[38] I. M. Yaglom, *Geometric Transformations III*, New Mathematical Library, Vol. 24, MAA, Washington, DC, 1973.

Marcin Emil Kuczma graduated in 1966 from the University of Warsaw, Faculty of Mathematics. He has taught mathematics there since then, and his main research areas are real analysis, and measure theory. He is a member of the Polish Mathematical Society, and has served as a consultant on the advisory panel of the *Mathematics and Informatics Quarterly*. Marcin Kuczma has been active in the area of mathematical competitions, serving recently as a member of the Polish Mathematical Olympiad Committee, and of the World Federation of National Mathematics Competitions. The Federation honored him with the David Hilbert International Award in 1992. He has served as a Jury member at the International Mathematical Olympiad, the Austrian-Polish Mathematical Competition, the Czech-Polish-Slovak Mathematical Competition, and the Baltic-Way Mathematical Competition. Kuczma has served four times on the IMO Problems Selection Committee.

His publications include: *On Numerical Series* (in Polish); *Problems: 144 Problems of the Austrian-Polish Mathematical Competition 1978–1993*; *Polish and Austrian Mathematical Olympiads 1981–1985* (with Erich Windischbacher); and *Mathematical Olympiads* (in Polish). He has contributed many problems to various journals: *Crux Mathematicorum*, *American Mathematical Monthly*, *Journal of Recreational Mathematics*, and *Elemente der Mathematik*.

QA 43
,K776